Preface

These are the lecture notes of a one-year course on probability theory given at Princeton University in the 1967-68 academic year.

I regret that the notes cannot cover all the topics discussed in class. Specifically, the following topics are excluded: i) abstract Lebesgue measure space; ii) nonlinear prediction theory; iii) finite dimensional approximation to white noise.

Part I presents some elementary material, which will serve as introduction and background for our approach. Parts II and III, which are the main parts, deal with generalized stochastic processes, particularly white noise. The results of Part II and III are, of course, not our final goal, but they will, I believe, be quite useful for our aim, which will be explained in §0.

I wish to express my thanks to Professor M. Silverstein who read §§0-4 and §§6-10 of my manuscript to correct the wrong expressions and to give me valuable advice. I also enjoyed the help of Mr. M. Hitsuda who eliminated many inaccuracies.

Contents

STATIONARY STOCHASTIC PROCESSES

BY TAKEYUKI HIDA

Errata

↓ = count down from top; ↑ = count up from bottom

page	line			(should read)
6	↓ 2	$\mathfrak{O}_t$ with	→	$\mathfrak{O}'_t$ with
	↑ 2	$L^2(\Omega, \ )$	→	$L^2(\Omega, P)$
14	↑ 4	Together	→	together
18	↓ 9	$(\cdots)X_{n+1}(\ )+(\cdots)X_{n+1}(\ )$	→	$2^n(\cdots)X_{n+1}(\ )+2^n(\cdots)X_{n+1}(\ )$
21-22		H sould read Ⓗ		
24	↓ 6	$\lvert S_{m,n}(t,\omega)$	→	$\lvert S_{m,n}(t,\omega)\rvert$
27	(the right hand of formula (4) should read)			$c\sqrt{2\lvert t-t'\rvert \log \frac{1}{\lvert t-t'\rvert}}$
29	↓ 7	The,	→	Then,
30	↓ 4	$(< t < \infty$	→	$0 < t < \infty$
	↓ 6	$\lvert B(t,\omega)$	→	$\lvert B(t,\omega)\rvert$
31	↑ 5	$f(x)$	→	$F(x)$
36	↓ 8	(insert) $F^{o*}(x) = \delta_{x,o}$		
	↑ 9	$(\lambda t)^n$	→	λ^n
43	↓ 10	Z_k, Z_j	→	z_k, z_j
44	↓ 1	Z	→	z
45	↑ 3	page	→	page 38
47	(between ↓ 8 and ↓ 9 insert)			$L(t,\omega)$ we obtain another Lévy process
	↓ 9	$L(t,\omega) \equiv$	→	$(g_a L)(t,\omega) \equiv$
49	↓ 5	$0 \leqq t > \infty$	→	$0 \leqq t < \infty$
51	↓ 10	$\lvert \Pi(-\alpha)$	→	$\lvert \Gamma(-\alpha)\rvert$
74	↑ 3	p 40	→	p 38
101	↑ 3	R^1	→	C^1
103	↓ 6	(after $\mathcal{H}$, insert) up to const. $\sqrt{n!}$, replacing c_k with $\sqrt{2}^{-k_j}$ in (10).		
104	↑ 4	(Πk_j)	→	$(\Pi k_j!)$
109	↓ 7	ρ	→	σ

110	↓4, 9, 13	$L^2(R^n)$	⟶	$\widehat{L^2(R^n)}$
114	↓ 7	φ - 5	⟶	φ - t
117	↑ 9	{ , h, A}	⟶	$\{\mathscr{A}, h, A\}$
118	↓ 12	NAUNAsN	⟶	$NA \cup NAsN$
122	↓ 10	$\sigma = F_1$	⟶	$\sigma \sim F_1$
125	↓ 3	$e^{-1<x_0, \xi>}$	⟶	$e^{-i<x_0, \xi>}$
126	↓ 6	$\exp(-\frac{1}{2}<y, x_0>)$	⟶	$\exp(-\frac{i}{2}<y, x_0>)$
127	↓ 10	$g \quad O^*_\infty(E), \xi \quad E\}$	⟶	$g \in O^*_\infty(E), \xi \in E\}$
129	↑ 10	othronormal	⟶	orthonormal
130	↓ 1	$\exp[-\frac{1}{2}<x, \xi>t]$	⟶	$\exp[-\frac{i}{2}<x, \xi>t]$
131	↑ 7	page 114	⟶	page 124
133	↓ 4	$\xi / \\|\quad\\|$	⟶	$\xi / \\|\xi\\|$
	↓ 8	$g \quad O_\infty$	⟶	$g \in O_\infty$
136	↑ 5	$n \leqq 0$	⟶	$n \geqq 0$
	↑ 2	< + <	⟶	< t <
137	↓ 1	p. 102	⟶	p. 111
	↓ 7	$\psi^m_n(x) =$	⟶	$\psi^m_n(x) d\mu(x) =$
	↓ 8	$\int \cdots \int_{R^n}$	⟶	$n! \int \cdots \int_{R^n}$
139	↓ 1, 8, 11	$dB(t)_{t=0}$	⟶	$dB(t)\|_{t=0}$
142	↓ 9	$\tau \geqq t)$	⟶	$\tau \leqq t)$
151	↓ 11	Where	⟶	with
157	↑ 1, 3	a + 1	⟶	$\alpha + 1$
158	((17) should read) $\exp\{\int\int(e^{i\xi(t)u} - 1 - i\xi(t)u)\|u\|^{-(\alpha+1)} du\, dt\}$			
160	↓ 3	wubgroup	⟶	subgroup
A 3	↑ 1	$b_r(g)$	⟶	$b_r(y)$

STATIONARY STOCHASTIC PROCESSES

WHITE NOISE

Professor T. Hida

STATIONARY STOCHASTIC PROCESSES

WHITE NOISE

§.0. Introduction

The ideas presented in this course were inspired by certain investigations of stationary stochastic processes using nonlinear operators acting on them, e.g. nonlinear prediction theory and the discussions about innovations for a stationary process.

We shall be interested in functionals of certain basic or fundamental stationary stochastic processes. In the discrete parameter case, they are the familiar sequences of independent identically distributed random variables. In the continuous parameter case, which concerns us, they are stationary processes with independent values at every moment in the sense of Gelfand. A particularly important role will be played by the so-called (Gaussian) white noise and also a generalized white noise. Before studying such processes, we will need to treat some preliminary matters. For example, white noise is not a stochastic process in the usual sense. Therefore we shall be led to give a new definition for a stationary stochastic process.

Once our general set-up has been established, we proceed to the analysis on Hilbert space arising from a stationary process with independent values at every moment. A detailed discussion can be given only in the case of white noise, when most of the results come from the classical theory for Wiener space and from the theory of Brownian motion. The greater part of the classical theory of stationary processes can be handled in our scheme.

We are in a position to analyze a given stationary process $X(t)$. We are motivated by the following formal expression due to P. Lévy:

$$\delta X(t) = \Phi(X(\tau);\ \tau \leqq t,\ \xi_t,\ t)$$

where Φ is a certain functional and $\{\xi_t\}$ forms a system of independent random variables. This equation is hard to illustrate rigorously, but on an intuitive level it is full of meaning. For example $\{\xi_t\}$ should be regarded as a process with independent values at every moment and plays the role of the marginal random variables which measure the information at each moment. Following along the lines suggested by the above expression, we shall consider not only functionals of ξ_t but also the flows by which ξ_t is transformed into other processes of the same type. In particular, we shall study flows which will serve to characterize the type of the measure induced by the given stationary process with independent values at every moment. In this connection we shall find that our study is closely related to the representation theory of Lie groups.

Finally we shall discuss some applications to physics and engineering where infinite dimensional analysis is required.

Part I

§.1. Background

In this article we shall prepare some basic concepts as background for our discussions.

1.1 Probability space

Let Ω be a certain non-empty set. Each element ω of Ω is supposed to be an elementary event or a probability parameter. Let Ⓑ be a σ-field of subsets of Ω, i.e., Ⓑ satisfies

i) If Ⓑ $\ni A_n$ for $n = 1,2,\dots$, then $\bigcup_n A_n \in$ Ⓑ

ii) If Ⓑ $\ni A$, then $A^c \equiv \Omega - A \in$ Ⓑ

iii) Ⓑ $\ni \emptyset$ (empty set).

A set belonging to Ⓑ is called an <u>event</u> or a <u>measurable</u> set (w. r. t. Ⓑ). A (countably additive) measure P defined on Ⓑ is called a <u>probability measure</u> if $P(\Omega) = 1$. The number $P(A)$ is called the probability of the event A.

A set Ω with elements ω, together with a σ-field Ⓑ and a probability measure P on $\textcircled{B}$ constitutes a probability space. We denote it by $\Omega(\textcircled{B},P)$ or $(\Omega, \textcircled{B}, P)$ or simply by Ω.

If A_n for $n = 1,2,\ldots,$ belong to $\textcircled{B}$, the following sets again belong to $\textcircled{B}$:

$$\bigcap_{n=1}^{\infty} A_n,$$

$$\text{inferior limit of } A_n = \underline{\lim}\, A_n = \bigcup_n \bigcap_{k \geq n} A_k,$$

$$\text{superior limit of } A_n = \overline{\lim}\, A_n = \bigcap_n \bigcup_{k \geq n} A_k,$$

(We sometimes refer to $\overline{\lim}\, A_n$ as the event that the A_n's occur infinitely often).

1.2. Random variable and probability distribution.

A real-valued function $X(\omega)$ on Ω is called a random variable (r.v.) on the probability space $(\Omega, \textcircled{B}, P)$ if X is measurable w.r.t. the σ-field $\textcircled{B}$. Given a r.v. X, a probability measure P_X is naturally defined on the measurable space $(R', \textcircled{B}(R'))$ in the following manner:

$$P_X(B) = P(X^{-1}(B)) \qquad B \in \textcircled{B}(R').$$

where $R' = (-\infty, \infty)$ and $\textcircled{B}(R')$ is the usual Borel field. Thus defined, P_X is called the probability distribution of the r.v. X.

In a similar manner we can define a complex-valued or vector-valued r.v. and its probability distribution.

Remarks. We will also give more general definitions of a r.v. and its probability distribution, which play important roles in the later articles.

If X is a real r.v., the expectation or mean of X is the integral

$$E(X) = \int X(\omega)P(d\omega) = \int x\, P_X(d\, x)$$

provided X is integrable. The n-th moment and the n-th absolute moment of X are the expectations of X^n and of $|X|^n$ respectively. $E[X - E(X)]^2$ is called the variance of the r.v. X.

1.3. Sequence of Events and r.v.'s.

We are interested in sequences of events and of r.v.'s which are independent.

i) Events A_t, (where t runs over a not necessarily finite set

T) to be independent if, for any finite subset $(t_1, \dots, t_n)$ of T,

$$P(\bigcap_{k=1}^{n} A_{t_k}) = \prod_{k=1}^{n} P(A_{t_k}).$$

ii) σ-fields $\mathfrak{G}_t$, (where $t \in T$ and $\mathfrak{C}_t \subset \mathfrak{B}$) are said to be independent if the events A_t are independent for any choice of $A_t \in \mathfrak{C}_t$.

If the σ-fields $\textcircled{C}_t$ are independent, then any sub-σ-fields $\textcircled{C}_t$ with $\textcircled{C}'_t \subset \textcircled{C}_t$ are also independent.

iii) r.v.'s X_t on the same probability space are said to be independent if the $\textcircled{B}(X_t)$ are independent, where $\textcircled{B}(X_t)$ is the smallest σ-field with respect to which X_t is measurable.

If r.v.'s X_t are independent, we can easily prove the following equations:

(1) $$P(X_{t_k} \in B_{t_k}, k = 1,2,\ldots,n) = \prod_{k=1}^{n} P(X_{t_k} \in B_{t_k})$$

for any finite subset $(t_1,\ldots,t_n)$ and for any B_{t_k}'s in $\textcircled{B}(R')$.

(2) $$E(\prod_{k=1}^{n} X_{t_k}) = \prod_{k=1}^{n} E(X_{t_k})$$

for any finite subset $(t_1,\ldots,t_n)$ provided $E(|X_{t_k}|) < \infty$ $k = =1,2,\ldots,n$.

For a system of independent r.v.'s with finite variances we can give a geometric representation, since each r.v. can be regarded as an element of the Hilbert space $L^2(\Omega,P)$. The mean and the second order absolute moment of a r.v. X correspond to the inner product $(X,1)$ and the square of the norm $\|X\|^2$ in $L^2(\Omega,P)$ respectively. Independent r.v.'s with zero mean are orthogonal to each other in $L^2(\Omega,)$. If a sequence X_n converges to X in $L^2(\Omega,P)$, we say that X_n converges to X in the <u>mean-square</u> or <u>m.q.</u> (moyenne quadratique).

Example. Let $\{X_n, n = 1,2,\ldots\}$ be a Gaussian system, i.e., the joint distribution of any finite collection of the X_n's is Gaussian. Assume that $E(X_n) = 0$ for every n. Since X_n has finite variance, $X_n \in L^2(\Omega,P)$. Applying the Schmidt's orthogonalization we obtain the following representation:

$$X_n = \sum_{k=1}^{n} A_{n,k} \xi_k \tag{3}$$

Where $\{\xi_k\}$ is an ortho-normal system in $L^2(\Omega,P)$. It is easy to see that $\{\xi_k\}$ forms a Gaussian system. We note that orthogonality is equivalent to independence for Gaussian r.v.'s. Therefore $\{\xi_k\}$ must be a system of independent Gaussian random variables with the common distribution $N(0,1)$ (Gaussian distribution with mean 0 and variance 1).

Having obtained the representation (3), we can easily discuss the prediction theory and related questions for $\{X_n\}$. This suggests to us the investigation of P. Lévy's representation of a Gaussian process.

Next we consider a countable system of events $\{A_n\}$.

Theorem 1.1. (Borel-Cantelli)

I. If $\sum_n P(A_n) < \infty$, then $P(\overline{\lim_n} A_n) = 0$.

II. If the A_n are independent, $P(\overline{\lim_n} A_n) = 0$ or 1 according as $\sum_n P(A_n) < \infty$ or $= +\infty$.

<u>Proof</u>. I is easily proved.

For the proof of II let us assume that $\sum_n P(A_n) = +\infty$.

$$P(\overline{\lim_n} A_n) = \lim_{m\to\infty} \lim_{n\to\infty} P(\bigcup_{k=m}^{n} A_k)$$

$$= \lim_m \lim_n (1 - \prod_m^n P(A_k^c))$$

$$= \lim_m \lim_n (1 - \prod_m^n (1 - P(A_k))).$$

Our assumption implies that the product in the above expression approaches 0 as $n \to \infty$. Therefore

$$P(\overline{\lim_n} A_n) = 1.$$

The rest of the proof is the same as I.

If the A_n are independent and if, in particular, $\lim_n A_n = A$ exists, then $P(A)$ must be 0 or 1.

As an application of I we have the following proposition: if there exist sequences $\epsilon_n \geq 0$ and $\eta_n \geq 0$ such that $\sum_n \epsilon_n < \infty$ and $\sum_n \eta_n < \infty$, and if $P(|X_{n+1} - X_n| > \eta_n) < \epsilon_n$ holds for every n, then $\lim_n X_n(\omega)$ exists for almost all ω.

For a sequence of r.v.'s X_n on $(\Omega, \textcircled{B}, P)$ we introduce the σ-field $\textcircled{B}_n = B(X_n, X_{n+1}, \ldots)$, the smallest σ-field with respect to which all of the X_m with $m \geq n$ are measurable. Since the $\textcircled{B}_n$ decrease as n increases, the limit

$$\textcircled{C} = \lim_{n \to \infty} \textcircled{B}_n = \cap \textcircled{B}_n$$

exists and is again a σ-field. $\textcircled{C}$ is called the <u>tail</u> σ-field of the sequence X_n. A function is said to be a <u>tail function</u> if it is measurable with respect to $\textcircled{C}$.

<u>Examples</u> of tail functions

a) $\overline{\lim_n}\ X_n(\omega)$, $\quad \lim_n X_n(\omega)$,

b) $f(\omega) = \begin{cases} 1 & \text{for } \omega \text{ such that } \sum_n X_n(\omega) \text{ converges,} \\ 0 & \text{otherwise.} \end{cases}$

<u>Theorem</u> 1.2. (Zero-One Law)

Let X_n, $n = 1,2,\ldots$, be a sequence of independent r.v.'s. Then the tail σ-field is equivalent to the trivial field $\{\emptyset, \Omega\}$, i.e. each member of the tail σ-field has measure 0 or 1.

<u>Proof</u>. The tail σ-field $\textcircled{C}$, being a sub-σ-field of $\textcircled{B}(X_{n+1}, X_{n+2}, \ldots)$, is independent of $\textcircled{B}(X_1, \ldots, X_n)$. Since this is true for every n, also $\textcircled{C}$ is independent of $\textcircled{B}(X_1, X_2, \ldots)$, which includes $\textcircled{C}$. Hence $\textcircled{C}$ is independent of itself. Thus, for any event A in $\textcircled{C}$,

$$P(A) = P(A \cap A) = P(A) \cdot P(A),$$

and therefore $P(A) = 0$ or 1.

<u>Corollary</u>. Let X_n be a sequence of independent r.v.'s. Then X_n converges a.s. or diverges a.s., and similarly for the sum $\sum_n X_n$.

Two sequences X_n and Y_n defined on the same probability space are called tail equivalent if they differ a.s. only by a finite number of terms. If $\sum_n P(X_n \neq Y_n) < \infty$, then the sequences are tail equivalent (Borel-Cantelli theorem I). In this case the series ΣX_n and ΣY_n are convergence equivalent, i.e., the sets on which they converge differ by a set of measure 0. Thus, so far as questions of convergence are concerned, these series can be used interchangably.

1.4. Law of large numbers.

We begin with two estimates of probabilities in terms of moments.

i) Tchebichev inequality. Let $g(x)$ be an even non negative function which is positive and non-decreasing on $(0, \infty)$. Then for any positive ε, we have

$$P(|X| \geq \varepsilon) \leqq \frac{E\{g(X)\}}{g(\varepsilon)}$$

ii) Kolmogrov inequality. Let X_n be a sequence of independent r.v.'s with finite means. Set $S_n = \sum_1^n X_k$. Then for any positive ε, we have

$$P(\max_{k \leq n} |S_k - E(S_k)| \geq \varepsilon) \leq \frac{1}{\varepsilon^2} V\{S_n\},$$

where V denotes the variance.

Tchebichev inequality follows from the elementary computation

$$E\{g(X)\} \geq E\{g(X); |X| \geq \varepsilon\} \geq g(\varepsilon) \quad P[|X| \geq \varepsilon].$$

Kolmogorov inequality is more subtle. To prove it, we assume that $E(X_k) = 0$ for all k and we set

$$M_k = \max_{j \leq k} |S_j|$$

$$A_k = (M_k < \varepsilon)$$

$$B_k = (M_{k-1} < \varepsilon, \ |S_k| \geq \varepsilon).$$

Then we have

$$B_k = A_{k-1} - A_k, \quad \text{and}$$

$$A_n^c = \sum_{k=1}^{n} B_k \quad \text{(direct sum)}.$$

Consider the integral

$$\int_{B_k} S_n^2(\omega)dP(\omega) = \int S_n^2 I_{B_k} \, dP,$$ (I_B is the indicator function of B)

$$= \int [(S_k I_{B_k})^2 + (S_n - S_k)^2 I_{B_k}^2] dP(\omega)$$

$$\geq E(S_k I_{B_k})^2 \geq \varepsilon^2 P(B_k)$$

(The crucial third step uses the independence of $S_k I_{B_k}$ and $S_n - S_k$.)

Summing over $k = 1,2,\ldots,n,$ we have

$$V(S_n) = E\,S_n^2 \geq \int_{A_n^c} S_n^2 \, dP(\omega) = \sum_{k=1}^{n} \int_{B_k} S_n^2 \, dP(\omega)$$

$$\geq \varepsilon^2 \sum_k P(B_k) = \varepsilon^2 \, P(A_n^c).$$

Suppose that the X_n are independent and identically distributed with finite variance. Then Tchebichev inequality gives

$$\lim_{n \to \infty} P\left(\left|\frac{1}{n} \sum_{k=1}^{n} X_k - m\right| > \varepsilon\right) = 0,$$

where m is the mean of X_n. Kolmogorov inequality yields the following stronger result:

<u>Theorem 1.3</u>. If the r.v.'s X_n are independent and have finite variances, then $\sum \frac{V(X_n)}{b_n^2} < \infty$ with increasing positive b_n implies that

$$\frac{1}{b_n} (S_n - E(S_n)) \longrightarrow 0 \quad \text{a.s.}$$

In particular we have the following classical form of the <u>strong law of large numbers</u>.

<u>Corollary</u>. If the independent r.v.'s X_n are identically distributed with finite mean m and finite variance, then

$$\frac{1}{n} S_n \longrightarrow m \quad \text{a.s.}$$

The corollary is still true if we drop the assumption of

finite variance, but an additional argument is required.

[Bibliography]

[1] W. Feller, An introduction to probability theory and its application. I,II. Wiley, 1950, 1966.

[2] P. R. Halmos, Measure theory. Van Nostrand, 1950.

[3] M. Loéve, Probability theory, 3rd ed. Van Nostrand, 1963.

§.2. Brownian motion.

First we shall define a standard Brownian motion and give one possible construction for it. Then we shall see that Brownian motion determines a measure on function space.

2.1. Definition of Brownian motion

A system of r.v.'s $B(t,\omega)$, $\omega \in \Omega$, $0 \le t < \infty$, is called a standard Brownian motion if it satisfies the following conditions:

i) The system $\{B(t);\ 0 \le t < \infty\}$ is Gaussian

ii) $B(0) = 0$, a.s..

iii) The probability distribution of $B(t) - B(s)$ is $N(0, |t-s|)$.

From iii) it follows that $E(B(t)) = 0$ for every t. Also for $u < s < t$, the relation

$$V(B(t)-B(u)) = V(B(t)-B(s)) + V(B(s)-B(u)) + 2E\{(B(t)-B(s))(B(s)-B(u))\}$$

Together with condition iii) shows that $B(t) - B(s)$ and $B(s) - B(u)$ are independent. Thus $E(B(t)\,B(s)) = \min(t,s)$. Similarly, it can be shown that $\{B(t_{i+1}) - B(t_i);\ t_1 < t_2 < \dots < t$ is a system of independent Gaussian r.v.'s and therefore

(1) $B(t) - B(s)$ with $s < t$ is independent of the system $\{B(u); u \leq s\}$. In other words, $B(t)$ is an <u>additive</u> stochastic process.

<u>Remarks</u>. Property (1) suggests that $\frac{d}{dt} X(t)$, if it exists in a suitable sense, is a system of independent r.v.'s. We will discuss this in §.5.

The conditions i), ii), iii) determine the probability distribution of $(B(t_1), \ldots, B(t_n))$ for any finite subset $(t_1, \ldots, t_n)$ of $[0, \infty)$. The system of these distributions satisfies the necessary consistency conditions for the Kolmogorov extension theorem; this assures the existence of a standard Brownian motion with $\Omega = R^{[0, \infty)}$. However we prefer to give a direct construction in order to clarify certain important properties.

We first consider a geometric representation for Brownian motion. Let M(resp.M_t) be a subspace $L^2 \equiv L^2(\Omega, P)$ spanned by the $B(t)$, $0 \leq t < \infty$ (resp. the $B(s)$, $0 \leq s \leq t$). Since the norm $\|B(t) - B(s)\|$ in L^2 is $\sqrt{|t - s|}$, the random variables $B(t)$ form a continuous curve in L^2, i.e. $B(t)$ is mean (square) continuous in t. Therefore the subspace M is separable. The property (1) indicates that the $B(t)$ "looks like" the diagonal line in the space M.

Consider for $t_1 < t < t_2$ the conditional expectation $E(B(t)/B(t_1), B(t_2))$. As was pointed out in the Example of § 1.3, $E(B(t)/B(t_1), B(t_2))$ is the projection of $B(t)$ on the plane spanned by $B(t_1)$ and $B(t_2)$ in M. In fact,

$$B(t) = E(B(t)/B(t_1), B(t_2)) + \sigma(t)\,\xi(t), \tag{2}$$

where $\xi(t)$ is independent of $(B(t_1), B(t_2))$ and has distribution $N(0,1)$, and where

$$E(B(t)/B(t_1), B(t_2)) = \frac{t_2 - t}{t_2 - t_1} B(t_1) + \frac{t - t_1}{t_2 - t_1} B(t_2)$$

$$\sigma(t) = \sqrt{(t - t_1)(t_2 - t)/(t_2 - t_1)}$$

Speaking in geometric terms, we say that the conditional expectation of $B(t)$ is obtained by linear interpolation of the vectors $B(t_1)$ and $B(t_2)$.

It should be noted that $\xi(t)$ is independent of $\{B(s);\ s \notin (t_1,t_2)\}$. Thus if we take two nonoverlapping intervals, say $[t_1,t_2]$, $[s_1,s_2]$ with $t_2 < s_1$, then $\{\xi(t);\ t_1 < t < t_2\}$ and $\{\xi(s);\ s_1 < s < s_2\}$ are independent. This suggests our construction of Brownian motion. Before proceeding to that construction, we insert an interesting remark concerning the Gaussian process $\xi(t)$ for $t_1 \leq t \leq t_2$.

Consider the covariance function $\rho(t,s) = E(\xi(t)\xi(s))$, $t_1 \leq t < s \leq t_2$. From (2) there follows

$$\rho(t,s) = \sqrt{\frac{(t - t_1)(t_2 - s)}{(s - t_1)(t_2 - t)}}$$

But $\rho^2(t,s)$ is nothing but the anharmonic ratio of the four numbers t_1, t_2, t, s, and therefore ρ is invariant under any projective transformation acting on the t-axis. Since the law for a Gaussian system with zero mean is determined uniquely by its covariance function, we conclude that the law for $\xi(t)$ is invariant under any projective transformation provided that the time interval $[0, \infty)$ is carried into itself.

As a special case, we note that $\frac{B(t)}{\sqrt{t}}$, $0 < t < \infty$ and $\sqrt{t}\ B(\frac{1}{t})$, $0 < t < \infty$, have the same law.

2.2. Construction of Brownian motion

For simplicity we shall construct Brownian motion with the time parameter space $T = [0, 1]$. The case $T = [0, \infty)$ requires only a slight modification.

We begin with a countable system of independent Gaussian r.v.'s $\xi_n(\omega)$, $n = 1,2,\ldots$, $\omega \in \Omega\ (\mathcal{B}, P)$, with the common distribution $N(0, 1)$. Let T_n, $n \geq 1$, be the set of binary points of the form $k/2^{n-1}$, $k = 0, 1, \ldots, 2^{n-1}$, and let $T_0 = \bigcup_n T_n$.

For $t \in T_1$, i.e. $t = 0$ or 1, let $X_1(t)$ be defined by

$$X_1(0) = 0$$

$$X_1(1) = \xi_1$$

$$X_1(t) = t\,\xi_1,\ t \notin T_1.$$

Suppose $X_n(t)$ has been defined for $0 \leq t \leq 1$. Then $X_{n+1}(t)$ is defined as follows:

$$X_{n+1}(t) = \begin{cases} X_n(t) & \text{for} \quad t \in T_n, \\ \frac{1}{2}\{X_n(t + \frac{1}{2^n}) + X_n(t - \frac{1}{2^n})\} + \frac{1}{2^n}\xi_k & \text{for} \quad t \in T_{n+1} - T_n, \\ (\frac{k+1}{2^n} - t)\, X_{n+1}(\frac{k}{2^n}) + (t - \frac{k}{2^n}) X_{n+1}(\frac{k+1}{2^n}) & \text{for} \quad \frac{k}{2^n} < t < \frac{k+1}{2^n}, \end{cases}$$

where for the second alternative, $k = k(t) = 2^{n-1} + \dfrac{2^n\,t + 1}{2}$.

Thus we have a sequence of Gaussian processes $\{X_n(t),\ t \in [0,1]\}$, $n = 1,2,$

It is easy to check by induction that $B(t)$, $t \in T_n$, and $X_n(t)$, $t \in T_n$, have the same distribution. We expect therefore that $X_n(t)$ approaches Brownian motion in a suitable sense, which is indeed the case. If $t \in T_o = \bigcup_n T_n$, then the limit of the sequence $X_n(t)$ exists in $L^2(\Omega, P)$ since all the $X_n(t)$ are the same for $n \geq N = N(t)$. Let $X(t)$ denote the $\lim_{n \to \infty} X_n(t)$. Then it is easy to see that $X(t)$ is uniformly continuous (w.r.t. the norm of L^2) for $t \in T_o$. Hence $X(t)$ $t \in T_o$, can be extended to a continuous curve $X(t)$, $0 \leq t \leq 1$, in L^2. This is again a Gaussian system and

$$E\,X(t) \equiv 0$$

$$V(X(t) - X(s)) = |t - s|;$$

thus $X(t)$, $0 \leq t \leq 1$, is a realization of Brownian motion. However this argument yields no information about the behavior of individual sample functions $X(\cdot,\omega)$. Therefore we use an alternative approach. We start from the Gaussian systems $X_n(t,\omega)$, $n = 1,2,\ldots$ and prove that the limit $\bar{X}(t,\omega) = \lim_{n\to\infty} X_n(t,\omega)$ exists a.s. and that $\bar{X}(t,\omega)$, $0 \leq t \leq 1$, is continuous in t for almost all ω. To do this we set

$$Y_n(t,\omega) = X_{n+1}(t,\omega) - X_n(t,\omega)$$

Then we have

$$Y_n(t) = 0, \quad \text{if} \quad t \in T_n$$

and

$$\max_{0 \leq t \leq 1} |Y_n(t,\omega)| = \max_{2^{n-1} \leq k \leq 2^n} \frac{1}{2^n} |\xi_k(\omega)|.$$

Noting the inequality

$$p_n = P(\max_{0 \leq t \leq 1} |Y_n(t)| > \lambda_n) \leqq 2^{n-1} P(|\xi_k| \geq 2^n \lambda_n)$$

$$\leqq 2^{n-1} \sqrt{\frac{2}{\pi}} \frac{1}{2^n \lambda_n} e^{-\frac{1}{2}(2^n \lambda_n)^2}$$

we see that $\sum_n p_n$ converges if we put $\lambda_n = \sqrt{2 C n \log 2} \cdot 2^{-n}$, with $C > 1$. Applying the Borel-Cantelli theorem, we have, except for finitely many n,

$$\max |Y_n(t,\omega)| \leqq \sqrt{2 C n \log 2} \cdot 2^{-n} \qquad \text{a.s.,}$$

which proves our assertion. Since the convergence is uniform in t, the sample functions $\bar{X}(t.\omega)$ are continuous in t for almost all ω.

It is easy to see that this limit $\bar{X}(t,\omega)$ is equal to $X(t,\omega)$ a.s. for any fixed t.

Remark. The above construction is due to P. Lévy [5].

As an application of Levy's construction of Brownian motion we can prove the following

Proposition. Let $B(t,\omega)$, $0 \leq t \leq 1$, $\omega \in \Omega$, be a Brownian motion.

Then the probability space $(\Omega, \textcircled{B}, P)$ with $\textcircled{B} = B(B(t), 0 \leq t \leq 1)$ can be regarded as an <u>abstract Lebesque space</u> without atom. (c.f. V.A. Rokhlin [7])

2.3. <u>Random Fourier series</u>. (Paley-Wiener [6]). Consider the subspace M_π of $L^2(\Omega, P)$ which is spanned by the $B(t), 0 \leq t \leq \pi$. For $\Delta = (s, t]$ denote the difference $B(t) - B(s)$ by $B(\Delta)$. To any step function $f(t) = \Sigma C_k I_{\Delta_k}$, Δ_k disjoint, we associate a r.v. $I(f) = \Sigma C_k B(\Delta_k) \in M_\pi$. The r.v. $I(f)$ does not depend on the representation of the function f. Thus we have defined a mapping I from the set of step functions defined on $[0, \pi]$ to a linear subspace of M_π. The mapping I has the following properties

i) I is linear, i.e.

$$I(af + bg) = aI(f) + bI(g), \quad a, b \text{ constants}$$

ii) $E(I(f)) = 0$

iii) $(I(f), I(g))_{M_\pi} = (f,g)_H, \quad \|I(f)\|_{M_\pi} = \|f\|_H$

where $\textcircled{H}$ is the Hilbert space $L^2([0, \pi], m)$, m Lebesque measure.

Therefore we can extend the mapping I to a unitary operator from $\textcircled{H}$ onto M_π so that i), ii), iii) remain true. The r.v.'s $I(f)$, $f \in H$, thus defined, are called <u>Wiener integrals</u>.

We sometimes write $I(f)$ in the form

$$\int_0^\pi f(t)dB(t)$$

since $B(\Delta)$ plays the role of a (random) measure although $B(t, \omega)$ is never of bounded variation.

We note further that the Wiener integral can be generalized to the case where $B(t)$ is an additive Gaussian process $X(t)$, $t \in T$ (interval), with $E\,X(t) = 0$. There H should be replaced by $L^2(T, \mu)$, where $\mu(\Delta) = V(X(\Delta))$.

Having established the isomorphism between Ⓗ and M_π, we can discuss the expansion of an element in M_π into a Fourier series. Let φ_n be a c.o.n.s. (complete orthonormal system) in H. Then the $I(\varphi_n)$ form a c.o.n.s. in M_π. Indeed, they form a system of independent r.v.'s with common distribution $N(0,1)$.

Corresponding to the expansion of f in Ⓗ

$$f = \sum_n a_n \varphi_n \qquad \text{with} \qquad a_n = (f, \varphi_n)_{Ⓗ}$$

we have

$$I(f) = \sum_n a_n I(\varphi_n).$$

In particular, if we take a c.o.n.s.

$$\varphi_0 = 1/\sqrt{\pi}$$

$$\varphi_n = \sqrt{2/\pi} \quad \cos nt, \ n = 1,2,\ldots$$

and take $f = I_{[0,t]}(I(f) = B(t))$, then we have

$$B(t) = \frac{t}{\sqrt{\pi}} X_0 + \sum_{n=1}^{\infty} \sqrt{\frac{2}{\pi}} \frac{\sin nt}{n} X_n, \tag{1}$$

where $X_n = I(\varphi_n)$.

The covergence of the series in (1) should be understood as the convergence in the mean, i.e. in M_π. We can also discuss almost sure convergence of the series in (1), which suggests that we construct Brownian motion by random Fourier series.

As in §.2.2, we begin with a countable system of independent Gaussian r.v.'s $\xi_n(\omega)$, $n = 0, 1, \ldots$, $\omega \in \Omega(Ⓑ, P)$, with the common distribution $N(0,1)$. Then we have

<u>Theorem</u> 2.1. The series

$$\frac{t}{\sqrt{\pi}} \xi_0(\omega) + \sum_{n=0}^{\infty} \sum_{k=2^n+1}^{2^{n+1}} \sqrt{\frac{2}{\pi}} \frac{\sin kt}{k} \xi_k(\omega), \quad 0 \le t \le \pi, \tag{2}$$

converges uniformly in t a.s., and the limit $B(t,\omega)$, $0 \leq t \leq \pi$, is a Brownian motion.

Proof. We set

$$S_{m,n}(t,\omega) = \sum_{k=m+1}^{n} \frac{\sin kt}{k} \xi_k(\omega),$$

and

$$T_{m,n}(\omega) = \sup_{0 \leq t \leq \pi} |S_{m,n}(t,\omega)$$

In order to show

$$\sum_n T_{2^n,2^{n+1}}(\omega) < \infty \qquad \text{a.s.},$$

it is enough to prove

$$(3) \qquad \sum_n E(T_{2^n,2^{n+1}}) < \infty .$$

From the inequality

$$|S_{m,n}(t)|^2 \leqq |\sum_{m+1}^{n} \frac{e^{ikt}}{k} \xi_k|^2 \leqq \sum_{m+n}^{n} \frac{1}{k^2} \xi_k^2 + 2 \sum_{p=1}^{n-m+1} |\sum_{h=m+1}^{n-p} \frac{\xi_h \xi_{h+p}}{h(h+p)}$$

we have

$$T_{m,n})\}^2 \leqq E(T^2_{m,n}) = E\{\sup_t |S_{m,n}(t)|^2\} \leqq \sum_{m+n}^{n} \frac{1}{k^2} + 2 \sum_{p=1}^{n-m+1} E(|\sum_{h=m+1}^{n-p} \frac{\xi_h \xi_{h+p}}{h(h+p)}|).$$

On the other hand it follows from our assumptions on the ξ_n that

$$\{E(|\sum_{h=m+1}^{n-p} \frac{\xi_h \xi_{h+p}}{h(h+p)}|)\}^2 \leqq E\{(\sum_{m+1}^{n-p} \frac{\xi_h \xi_{h+p}}{h(h+p)})^2\} = \sum_{h=m+1}^{n-p} \frac{E(\xi_h^2)E(\xi_{h+p}^2)}{h^2(h+p)^2} +$$

$$+ 2 \sum_{m+1 \leq h < k \leq n-p} \frac{E(\xi_h \xi_{h+p} \xi_k \xi_{k+p})}{h\, k(h+p)(k+p)}$$

Hence we have

$$\{E(T_{m,n})\}^2 \leqq \sum_{m+1}^{n} \frac{1}{k^2} + 2 \sum_{p=1}^{n-m-1} (\sum_{h=m+1}^{n-p} \frac{1}{h^2(h+p)^2})^{1/2}$$

$$\leqq (n-m)/m^2 + 2(n-m) \sqrt{(n-m)/m^4}\,.$$

Putting $n = 2m$, we obtain

$$E(T_{m,2m}) \leqq \sqrt{3}\, m^{-\frac{1}{4}},$$

which implies

$$\sum_{n=1}^{\infty} E(T_{2^n,2^{n+1}}) \leqq \sqrt{3} \sum_{n=1}^{\infty} 2^{-\frac{n}{4}} < \infty .$$

Thus we have proved that the series (2) converges absolutely and uniformly in t a.s. It follows that the $B(t,\omega)$ defined by (2) is continuous in t a.s.

We show next that the $B(t,\omega)$, $0 \leqq t \leqq \pi$, $\omega \in \Omega$, is a Brownian motion. We have already noted that the series (1) converges in M_π and therefore we can say the same for (2), the limit of which is equal to $B(t,\omega)$ a.s. for every t. This implies that the $B(t,\omega)$, $0 \leqq t \leqq \pi$, form a Gaussian system. Finally, it is easy to see that

$$E(B(t,\omega)) = 0$$

$$E(B(t,\omega)B(s,\omega)) = \frac{ts}{\pi} + \sum_{n=1}^{\infty} \frac{2}{\pi} \frac{\sin nt}{n} \frac{\sin ns}{n} = \min(t,s).$$

(Use the eigenfunction expansion for the kernel $\frac{ts}{\pi} - \min(t,s)$.)

2.4. Continuity of Sample Functions. Let $B(t,\omega)$, $0 \le t \le 1$, $\omega \in \Omega$, be Brownian motion. We prove

Theorem 2.2. Let $c > 1$, be fixed. For almost all ω there exists $1 > \delta = \delta(\omega) > 0$ such that

$$(4) \qquad |B(t',\omega) - B(t,\omega)| \leqq c \sqrt{2|t-t'| \quad \frac{1}{|t-t'|}}$$

whenever $|t'-t| < \delta$.

<u>Proof</u>. i) Let $h = 2^{-n}$. Since $B(t+h) - B(t)$ has a distribution $N(0,h)$, we have

$$\alpha_n = P(|B(t+h) - B(t)| > c\sqrt{2h \log h^{-1}}\ ) \leqq \frac{1}{c\sqrt{n\pi \log 2}}\, 2^{-c^2 n}.$$

Hence

$$P(\exists\, k, |B((k+1)h) - B(kh)| > c\sqrt{2h \log h^{-1}}\ ) \leqq \sum_n 2^n \alpha_n < \infty.$$

By Borel-Cantelli theorem, there exists $n_0 = n_0(\omega)$ such that for $n > n_0$

$$|B((k+1)h) - B(k\ h)| \leqq c\sqrt{2h \log h^{-1}}\ , \quad \text{a.s.}$$

holds for every $k \leqq 2^n$.

In a similar manner, we can prove that, for fixed $c > 0$ and N, there exists $n_1 = n_1(\omega)$ such that if $n > n_1(\omega)$

$$|B((k+\gamma)h) - B(k\,h)| \leqq c\sqrt{2\,\gamma h \log (\gamma h)^{-1}} \qquad \text{a.s.}$$

holds for every $\gamma \leqq N$ and every $k \leq 2^n$.

ii) Next we consider the case where

$$t = k\,2^{-n},\ t < t' < t + 2^{-n} \qquad \text{with} \qquad n > n_o.$$

By binary (base 2) decimal expansion

$$t' - t = \sum_{\nu=1}^{\infty} \varepsilon_\nu\, 2^{-n-\nu},\ \varepsilon_\nu = 0 \text{ or } 1,$$

and by using the result of i), we obtain

$$|B(t') - B(t)| \leqq c \sum_\nu \varepsilon_\nu \sqrt{2^{-n-\nu+1} \log 2^{n+\nu}} \leqq c'\sqrt{2^{-n-p+1} \log 2^{n+p}}$$

where $c' > 1$ and $p = \min\{\nu;\ \varepsilon_\nu \neq 0\}$. Hence

$$(5) \qquad |B(t') - B(t)| \leqq c'\sqrt{2|t'-t| \log \frac{1}{|t'-t|}}$$

holds a.s.

The inequality (5) is still true if t is of the form $k\cdot 2^{-n}$ and $t > t' > t - 2^{-n}$.

iii) As the final step we consider the case where both t' and t are arbitrary but we suppose that $t' > t$ and $t' - t$ is sufficiently small so that

$$N\, 2^{-n-1} < t' - t \leqq N\, 2^{-n}, \quad n > \mathrm{Max}(n_p, n_1)$$

hold. Then we can find t_1, t_1' and integers q, q' such that

$$q 2^{-n} < t \leqq t_1 = (q+1)2^{-n} < t_1' = q' 2^{-n} \leq t' < (q'+1)2^{-n}.$$

Let $c = 1 + 2\,\varepsilon$, $c' = 1 + \varepsilon$ and $N \geqq 16\, \dfrac{c'^2}{\varepsilon^2}$. The, by i), $n_o(\omega)$ and $n_1(\omega)$ are determined, and for $n > \max(n_o, n_1)$ we have

$$|B(t_1') - B(t_1)| < (1+\varepsilon) \sqrt{2|t'-t| \log \frac{1}{|t'-t|}}\,, \qquad \text{a.s.},$$

$$|B(t') - B(t_1')| + |B(t_1) - B(t)| < 2c' \sqrt{2^{n-1} \log 2^n}$$

$$< 2c' \sqrt{\frac{4|t'-t|}{N} \log \frac{N}{|t'-t|}}\,, \qquad \text{a.s.},$$

Now

$$|B(t') - B(t)| \leqq |B(t') - B(t_1')| + |B(t_1') - B(t_1)| + |B(t_1) - B(t)|,$$

and we may assume $|t'-t| < 1/N$. This establishes the inequality (4).

In particular, for $t \leq \delta$, we have

$$(6) \quad |B(t,\omega)| = |B(t,\omega) - B(0,\omega)| \leqq c\sqrt{2t \log t^{-1}}, \text{ a.s.}$$

Since $B(t)$ and $tB(\frac{1}{t})$, $(< t < \infty$, have the same law, it follows that there exists $T = T(\omega)$ such that for $t > T$

$$(7) \quad |B(t,\omega) \leqq c\sqrt{2t \log t}, \quad \text{a.s.}$$

which gives information about the asymptotic property of sample functi

Remark. Much finer results have been obtained both for asymptotic behavior and the modulus of continuity. Our results are rough approximations which will suffice for our later discussion.

[Bibliography]

[4] P. Lévy, Theorie de l'addition des variables aleatoires. Gauthier Villars. 1937.

[5] P. Lévy, Processus stochastiques et mouvement brownien. Gauthier Villars. 2e éd. 1965.

[6] R.E.A.C. Paley, N. Wiener, Fourier transforms in the complex domain. A.M.S. Colloq. Pub. 1934.

[7] V.A. Rokhlin, On the fundamental ideas of measure theory. A.M.S. Translations series 1 Vol. 10.

§3. Additive processes

We shall begin with very simple and elementary examples of additive processes, i.e. the Poisson process and compound Poisson processes, the sample functions of which are quite different from those of Brownian motion. Then we shall discuss, as a generalization of compound Poisson processes, Lévy processes with stationary increments.

As in the case of Brownian motion, a Lévy process determines a probability measure on function space. The tension group acting on the function space will serve to characterize stable processes. This leads to a probabilistic interpretation of Bochner's famous theory of subordination.

3.1. Poisson process

Let $\xi_n(\omega)$, $\omega \in \Omega$ $(\textcircled{B}, P)$, $n = 1, 2, \ldots$, be a system of independent r.v.'s with the same exponential distribution

$$(1) \qquad f(x) = P(\xi_n \leqq x) = \begin{cases} 1 - e^{-\lambda x}, & x \geqq 0 \\ 0, & x < 0 \end{cases}$$

where $\lambda > 0$. The expectation $E(\xi_n)$ is obviously $1/\lambda$.

<u>Lemma.</u> If $\xi_1, \ldots, \xi_n$ are mutually independent r.v.'s with the distribution (1), then the sum $S_n = \sum_{k=1}^{n} \xi_k$ has a density f_n and a distribution function $F_n (F_1 = F)$ given by

$$
(2) \qquad
\begin{aligned}
f_n(x) &= \begin{cases} \lambda \dfrac{(\lambda x)^{n-1}}{(n-1)!}\, e^{-\lambda x}, & x \geqq 0 \\[2ex] 0\,, & x < 0 \end{cases} \\[3ex]
F_n(x) &= \begin{cases} 1 - e^{\lambda x} \displaystyle\sum_{k=0}^{n-1} \dfrac{(\lambda x)^k}{k!}, & x \geqq 0 \\[2ex] 0\,, & x < 0 \end{cases}
\end{aligned}
$$

A <u>Poisson process</u> $P(t,\omega)$, $\omega \in \Omega\,(\textcircled{B},P)$, $0 \leqq t < \infty$, can be defined by

$$
(3) \qquad P(t,\omega) = \begin{cases} \operatorname{Max}\{k:\ S_k(\omega) \leqq t\}\,, & (S_0(\omega) = 0) \\ \infty\,, \text{ if } S_k \leq t & \text{for all } k \end{cases}
$$

We note that $P(t,\omega) = n$ if and only if $S_n(\omega) \leqq t$ and $S_{n+1}(\omega) > t$.

The following assertions follow directly from (3) and the above lemma.

1) For any t, $P(t,\omega)$ is well-defined (it is a r.v.) and it is finite a.s.

2) For almost all ω, $P(\cdot,\omega)$ is an integral-valued, increasing right continuous function.

3) The probability distribution of $P(t,\omega)$ is Poisson:

$$
(4) \qquad
\begin{aligned}
P(P(t,\omega) = n) &= \frac{(\lambda t)^n}{n!}\, e^{-\lambda t}, \quad n = 0,1,2,\ldots, \\
P(P(t,\omega) = \infty) &= 0\,.
\end{aligned}
$$

In particular $E(P(t,\omega)) = \lambda t$ and $V(P(t,\omega)) = \lambda t$.

<u>Theorem 3.1.</u> $P(t,\omega)$, $0 \leq t < \infty$, is an additive process with stationary increments.

The proof is easy if we note

$$P(\xi > a + x/\xi > a) = P(\xi > x)$$

for a r.v. ξ with the exponential distribution.

Theorem 3.1 and (4) identify $P(t,\omega)$, $0 \leq t < \infty$ as a Poisson process. We state some further properties of a Poisson process:

4) The process $P(t,\omega)$ has no fixed discontinuity:

$$P(\lim_{h \to 0} P(t+h,\omega) = P(t,\omega)) = 1 \quad \text{for every} \quad t > 0 .$$

5) $P(t,\omega)$ is continuous in the mean:

$$\lim_{h \to 0} E|P(t+h,\omega) - P(t,\omega)|^2 = 0 ,$$

and hence it is continuous in probability.

It is easy to obtain the characteristic function of $P(t,\omega)$

$$E(e^{izP(t,\omega)}) = \exp\{\lambda t(e^{iz} - 1)\} .$$

3.2 <u>Compound Poisson processes</u>

In this section we introduce compound Poisson processes. These are also additive processes with stationary increments and they may be regarded as generalizations of the ordinary Poisson process. We give two constructions for these processes. The first construction yields only a special case. However in Section 3.3 we shall generalize this construction to obtain not only the most general compound Poisson process but a much wider class of processes.

[A] Let $P_k(t,\omega)$, $k = 1,2,\ldots,n$ be a system of independent Poisson processes such that

$$E\, P_k(t,\omega) = \lambda_k t$$

and let u_k, $k = 1,2,\ldots,n$ be n different non-zero real numbers. Consider a process $X(t,\omega)$ defined by

$$X(t,\omega) = \sum_{k=1}^{n} u_k P_k(t,\omega) . \tag{5}$$

Then it is quite easy to see that

1) $X(t,\omega)$ is an additive process with stationary increments and $E\, X(t) = (\sum_k u_k \lambda_k) t$.

2) For almost all ω, the sample function $X(\cdot,\omega)$ increases or decreases only by jumps.

3) $X(t,\omega)$ has no fixed discontinuities.

4) $X(t,\omega)$ is continuous in the mean.

If we want to know the probability distribution of the $X(t,\omega)$, it is enough to know the characteristic function of $X(1)$, since $X(t,\omega)$ is additive and has stationary increments.

$$E(e^{izX(1)}) = \prod_k E(e^{izu_k P_k(1)}) = \exp\{\sum_k \lambda_k (e^{izu_k} - 1)\} . \tag{6}$$

Note that

$$P\,(\text{the first jump of } X(t,\omega) \text{ is } u_k \text{ for } t \geq t_0)$$

$$= P\,(\quad\text{"}\quad\quad\text{"}\quad \text{for } t \geq 0)$$

$$= \int_0^\infty e^{-\sum_{j \neq k} \lambda_j x} \lambda_k e^{-\lambda_k x} dx$$

$$= \lambda_k / \sum_{j=1}^n \lambda_j$$

which reveals one role of the parameter λ_j. Another role is revealed by

$$P(X(t,\omega) \text{ has no jump on the period } [s, s+\tau])$$

$$= \prod_k P(X_k(t,\omega) \text{ has no jump on } [0,\tau])$$

$$= \prod_k e^{-\lambda_k \tau} = e^{-\tau \Sigma \lambda_k} .$$

Remark. Even in the case where $X(t,\omega)$ is defined by the sum of infinitely many independent Poisson processes, the situation is nearly the same if $\sum_j \lambda_j < \infty$ and $\sum_j u_j^2 \lambda_j < \infty$. If $\sum_j \lambda_j = \infty$, we need some trick to show that the sum (5) converges. For the corresponding additive process, the sample function has jump discontinuities on any finite interval.

[B] We now give the second construction. Let $\eta_1(\omega), \eta_2(\omega), \ldots$ be independent r.v.'s with common distribution function F and let $P(t,\omega)$ be a Poisson process with parameter $\lambda > 0$ which is independent of the $\eta_n(\omega)$. We define $Y(t,\omega)$, the compound Poisson process, by

$$(7) \qquad Y(t,\omega) = \begin{cases} \eta_1(\omega) + \eta_2(\omega) + \ldots + \eta_{P(t,\omega)}(\omega) \, , & \text{if} \quad P(t,\omega) \geq 1 \\ 0 \, , \quad \text{if} \quad P(t,\omega) = 0 \, . \end{cases}$$

The sample function $Y(t,\omega)$ jumps by $\eta_n(\omega)$ at the instant when $P(t,\omega)$ jumps from $n-1$ to n.

The distribution function $G_t(x)$ of $Y(t)$ is easily obtained as follows:

$$(8) \qquad G_t(x) = P(Y(t,\omega) \leq x) = \sum_{n=0}^{\infty} \frac{(\lambda t)^n}{n!} e^{-\lambda t} F^{n*}(x) \, ,$$

where $F^{n*} = F^{(n-1)*} * F$ and $F^{1*} = F$.

The compound Poisson process is again an additive process with stationary increments. The characteristic function of $Y(1)$ is given by

$$(9) \qquad E(e^{iz\,Y(1)}) = \sum_{n=0}^{\infty} \frac{(\lambda t)^n}{n!} e^{-\lambda} \varphi(x)^n$$

$$= \exp\{\lambda \int (e^{izx} - 1) \, dF(x)\} \, .$$

Thus we see that the process $X(t,\omega)$ defined by (5) is the special case of a compound Poisson process, corresponding to F being concentrated on a finite set of non-zero integers. In particular, the ordinary Poisson process corresponds to F being concentrated at the single point 1.

3.3 Lévy processes

We proceed to define more general additive processes.

<u>Definition.</u> $X(t,\omega)$, $0 \leq t < \infty$, $\omega \in \Omega$, is said to be a <u>Lévy process</u> if it satisfies the following conditions:

i) $X(t,\omega)$ is additive

ii) $X(t,\omega)$ is continuous in probability, that is,

$$\lim_{t \to s} P(|X(t,\omega) - X(s,\omega)| > \varepsilon) = 0 \quad \text{for any } s ,$$

iii) for almost all ω the same function $X(t,\omega)$ is a <u>ruled</u> function, i.e. there always exist $X(t+0,\omega) = X(t,\omega)$ and $X(t-0,\omega)$ for every t .

The third condition is a question of selecting a regular version. That is, if a process $X(t)$ satisfies assumptions i), ii) we can form an additive process $X(t)$ which is equivalent to the given process $X(t)$ and satisfies in addition the condition iii). (J. L. Doob.)

We will be concerned only with Lévy processes having stationary increments. For simplicity we always assume that $X(0,\omega) = 0$ for every ω .

<u>Example 1.</u> The compound Poisson processes $X(t,\omega)$ and $Y(t,\omega)$ formed in §3.2 are all Lévy processes.

We now proceed to the construction of a Lévy process starting from a system of Poisson processes. Our construction is a generalization of the first construction given in §3.2. Note that since a Lévy process is additive and has no fixed discontinuity (by ii)), it suffices to form a Lévy process on the time interval $[0,1]$.

Let (P) = $\{P_I(t,\omega)\ ;\ 0 \leq t \leq 1$, I is an interval of the form (a,b] , a,b rational$\}$ be a system of Poisson processes $P_I(t,\omega)$ such that

a) $E(P_I(t,\omega)) = t \cdot n(I)$,

where n is a measure supported on $(-\infty,0) \cup (0,\infty)$ satisfying

$$\int_{|u|>0} \frac{u^2}{1+u^2}\, dn(u) < \infty \ , \tag{10}$$

b) if the I_n are disjoint, the P_{I_n} are independent,

c) if I is the sum of disjoint I_k's, $k = 1,2,\ldots,n$, then

$$P_I(t,\omega) = \sum_{k=1}^{n} P_{I_k}(t,\omega) \ .$$

Remark. The proof of the existence of such a system (P) is left to the reader.

Since $P_I(t,\omega)$ is additive in I, it looks like a measure. Indeed, the integral

$$S_I(t,\omega) = \int_I uP_{du}(t,\omega) \tag{11}$$

can be defined just as for the ordinary Rieman-Stieltjes integral. For I with $\bar{I}$ (the closure of I) $\not\ni 0$, set

$$S_m(t,\omega) = \sum_{i=1}^{m} i_{m_i} P_{I_{m_i}}(t,\omega)$$

where the I_{m_i} , $i = 1,2,\ldots,m$, are subintervals of I such that

$I = \Sigma I_{m_i}$, $I_{m_i} = (u_{m_{i-1}}, u_{m_i}]$ u_{m_i} rational $u_{m_i} - u_{m_{i-1}} = \frac{|I|}{m}$, $|I|$ the length of I. $S_m(t,\omega)$ is a Lévy process, and it is easy to see that

$$|S_m(t,\omega) - S_I(t,\omega)| \leq \frac{|I|}{m} P_I(1,\omega)$$

for every ω and every t, $0 \leq t \leq 1$. It follows that $S_I(t,\omega)$ is a Lévy process. We now prove

Theorem 3.2. The limit

$$(12) \qquad \lim_{p \to \infty} \int_{p > |u| > \frac{1}{p}} (uP_{du}(t,\omega) - \frac{tu}{1+u^2} dn(u)) \equiv X(t,\omega)$$

exists a.s. The convergence to the limit is uniform in t, $t \in [0,1]$. Moreover $X(t,\omega)$ is a Lévy process with stationary increments.

Proof. i) We already proved that $S_I(t,\omega)$ is a Lévy process. It is obvious that $S_I(t,\omega)$ has stationary increments. Therefore the process defined by

$$T_p(t,\omega) = \int_{p > |u| > \frac{1}{p}} (uP_{du}(t,\omega) - \frac{tu}{1+u^2} dn(u))$$

is again a Lévy process with stationary increments.

ii) We now proceed to prove that $T_p(t,\omega)$ converges to a Lévy process uniformly in t. To this end let us first establish the convergence of $A_p(t,\omega)$ given by the integral

$$A_p(t,\omega) = \int_1^p uP_{du}(t,\omega) .$$

Since $P_{(1,p]}(t,\omega)$ is non-negative and increasing in t,

$$\lim_{p\to\infty} E(P_{(1,p]}(t,\omega)) = \lim_{p} t\int_1^p dn(u) < \infty$$

proves the existence of the $\lim_{p\to\infty} P_{(1,p]}(t,\omega)$ a.s. Noting that $P_{(1,p]}(t,\omega)$ takes only integral values we can see that there exists an integer $p_o = p_o(\omega)$ such that

$$P_{(1,p]}(t,\omega) = P_{(1,p_o)}(t,\omega),\ p \geq p_o\ ,\ \text{for every}\ t\ .$$

This shows that

$$A_p(t,\omega) = A_q(t,\omega),\ p,q \geq p_o \quad \text{for every}\ t\ ,$$

which proves that $A_p(t,\omega)$ converges a.e. uniformly in t.

iii) The next step is the most difficult part. Set

$$X_k(t,\omega) = \int_{1/(k+1)}^{1/k} (uP_{du}(t,\omega) - t\ udn(u))\ ,\quad k=1,2,\ldots\ .$$

Then the mean and the variance of $X_k(t,\omega)$ are easily obtained:

$$E(X_k(t,\omega)) = 0$$

$$V(X_k(t,\omega)) = t\int_{1/(k+1)}^{1/k} u^2 dn(u)\ .$$

Since $\sum_k V(X_k(t,\omega)) < \infty$, and the $X_k(t,\omega)$ are mutually independent, the partial sum

$$B_p(t,\omega) = \sum_{k=1}^{p} X_k(t,\omega)$$

converges a.s. as $p \longrightarrow \infty$. More precisely $B_p(t,\omega)$ converges uniformly in t. To prove the uniform convergence we first obtain the following inequality. For a given small $\varepsilon > 0$, there exists $N = N(\varepsilon)$ such that

$$(13) \qquad P(\sup_t |B_m(t,\omega) - B_n(t,\omega)| > \varepsilon) < \varepsilon, \qquad m,n \geq N.$$

Using (13), we can prove that for any n

$$P(\max_{1 \leq k \leq n} \sup_t |B_{N+k}(t,\omega) - B_N(t,\omega)| > 2\varepsilon) < 2\varepsilon$$

by arguing as in the proof of the Kolmogorov inequality. Therefore we have

$$P(\sup_{m,n > N} \sup_t |B_n(t,\omega) - B_m(t,\omega)| > 4\varepsilon) < 4\varepsilon .$$

Letting $N \longrightarrow \infty$ and then $\varepsilon \longrightarrow 0$, we have

$$P(\lim_{N \to \infty} \sup_{m,n > N} \sup_t |B_n(t,\omega) - B_m(t,\omega)| > 0) = 0,$$

that is $B_p(t,\omega)$ converges uniformly in t for almost all ω as $p \longrightarrow \infty$.

Thus the $\lim_{p \to \infty} B_p(t,\omega)$ is a Lévy process with stationary increments.

iv) Finally, we consider the sum

$$T_p^+(t,\omega) = A_p(t,\omega) + B_p(t,\omega) - t \int_{\frac{1}{p}}^{p} \frac{u}{1+u^2}\, dn(u)$$

$$+ t \int_{\frac{1}{p}}^{1} \frac{u^3}{1+u^2}\, dn(u)$$

$$= \int_{p>u>\frac{1}{p}} (u\, P_{du}(t,\omega) - \frac{tu}{1+u^2}\, dn(u)) .$$

From the discussions in ii) and iii), we see that $T_p^+(t,\omega)$ converges uniformly in t for almost all ω as $p \longrightarrow \infty$. The same is true for $T_p^-(t,\omega) = T_p(t,\omega) - T_p^+(t,\omega)$. At the same time we see that the $\lim_{p\to\infty} T_p(t,\omega)$ is a Lévy process with stationary increments. Thus the theorem is proved.

The techniques used to prove Theorem 3.2 can also be used to compute the characteristic function of $X(t,\omega)$:

<u>Theorem 3.3.</u> If $X(t,\omega)$ is a Lévy process given by (12), then

$$(14) \qquad E(e^{iz\, X(t)}) = \exp\{t \int (e^{izu} - 1 - \frac{izu}{1+u^2}) dn(u)\} , \quad z \text{ real.}$$

<u>Proof.</u>

$$E(e^{izX(t)}) = \lim_{p\to\infty} E[\exp\{iz \int_{p>|u|>\frac{1}{p}} (uP_{du}(t,\omega) - \frac{tu}{1+u^2}\, dn(u)\}]$$

$$= \lim_{p\to\infty} [\exp\{-itz \int_{p>|u|>\frac{1}{p}} \frac{u}{1+u^2}\, dn(u)\}\cdot\exp\{t \int_{p>|u|>\frac{1}{p}} (e^{izu}-1) dn$$

(c.f. formulas (6), (7))

$$= \exp\{\int (e^{izu} - 1 - \frac{izu}{1+u^2}) dn(u)\} .$$

Here we note that the assumption (10) for $dn(u)$ is necessary. Also the characteristic function $\varphi(z) = E(e^{izX(1)})$ is the general form of the characteristic function of the <u>infinitely divisible law</u> missing the Gaussian part. The continuous function $\psi(z)$ with $\varphi(z) = \exp\{\psi(z)\}$ and $\psi(0) = 0$ is called the <u>ψ-function</u> corresponding to the Lévy process $X(t)$.

Once the ψ-function is given, we can find the joint distribution of $(X(t_1),\ldots,X(t_n))$ for any $t_1,\ldots,t_n$. In fact simple computations show that

$$E(\exp\{i \sum_{k=1}^{n} Z_k X(t_k)\}) = \exp\{\Sigma(t_k - t_{k-1})\psi(\sum_{j=k}^{n} Z_j)\} .$$

In this sense the distribution of the process $X(t,\omega)$ is uniquely determined by the ψ-function.

The measure $dn(u)$ is sometimes called the <u>Lévy measure</u> of the process. Here are some examples:

<u>Example 1.</u> To the Lévy process defined by (5) corresponds a discrete Lévy measure.

<u>Example 2.</u> The compound Poisson process $Y(t)$ defined by (7) has the Lévy measure $\lambda\, d\,F(u)$. Indeed $Y(t)$ is expressible in the form (12) if we subtract a term $t \cdot \int \frac{\lambda u}{1+u^2}\, d\,F(u)$ from $Y(t)$. This is an example having a finite Lévy measure.

<u>Example 3.</u> By Theorem 3.2 we see that a Lévy process with Lévy measure concentrated on $(0,\infty)$ and density $u^{-(1+\alpha)}$, $0 < \alpha < 1$,

can be formed. The Lévy process with $\psi(Z) = \int_0^\infty (e^{izu} - 1) \frac{1}{u^{1+\alpha}} du$ has increasing sample functions. They increase just by jumps, the size of which can be any positive number.

We come now to the relation between the Lévy measure and the jumps of sample functions.

Theorem 3.4. Let $X(t,\omega)$ be a Lévy process given by (12). Then we have, for any τ and s

$$P(X(t,\omega) \text{ has no jump on } [s,s+\tau]) = 0$$

if and only if

$$\int dn(u) = \infty .$$

Proof. By the definition of the integral of the form $\int_I uP_{du}(t,\omega)$ we can easily prove that

$$P(\int_{1/p}^{p} (uP_{du}(t,\omega) - tudn(u)) \text{ has no jump on } [s,s+\tau])$$

$$= e^{-\tau n((\frac{1}{p},p])} .$$

Our assertion follows upon letting $p \longrightarrow \infty$.

Let $X(t,\omega)$ be the Lévy process given by (12) and let $B(t,\omega)$ be a Brownian motion which is independent of $X(t,\omega)$. Then a process $L(t,\omega)$ defined by

$$L(t,\omega) = mt + \sigma B(t,\omega) + X(t,\omega) , \tag{15}$$

m,σ constants, is again a Lévy process with stationary increments. The characteristic function of $L(1)$ is

$$E(e^{izL(1)}) = \exp\{imz - \frac{\sigma^2}{2} z^2 + \psi(z)\} , \tag{16}$$

where ψ is the ψ-function corresponding to $X(t)$.

It should be noted that there is a sharp difference between the sample function of $B(t)$ and that of $X(t)$ although both $B(t)$ and $X(t)$ are Lévy processes with stationary increments. Indeed, for almost all ω, $B(t,\omega)$ is continuous while $X(t,\omega)$ increases or decreases only by jumps up to a linear function of t. Therefore the sample function of $L(t)$ can be decomposed into two parts, each having quite different continuity properties. We state this result formally in

Theorem 3.5. (Lévy decomposition theorem). Let $L(t,\omega)$, $0 \leq t \leq 1$ be a Lévy process with stationary increments. Then we can find constants m and σ, and we can form a Brownian motion $B(t,\omega)$ and a system of Poisson processes $\textcircled{P} = \{P_I(t,\omega)\}$ with the properties a), b), c) on page such that

$$L(t,\omega) = mt + \sigma B(t,\omega) + \lim_{p \to \infty} \int_{p > |u| > \frac{1}{p}} \left(uP_{du}(t,\omega) - \frac{tu}{1+u^2}\, dn(u)\right).$$

Moreover such a decomposition is unique.

For details of the proof of the theorem, see P. Lévy [11, Chapter V], K. Ito [9], [10]. We give an outline of the proof. First form a system of Poisson processes $P_I(t,\omega)$ by letting the jumps of $P_I(t,\omega)$ be the jumps of $L(t,\omega)$ with size $u \in I$. Then it is easy to check that the system $\textcircled{P} = \{P_I(t,\omega)\}$ satisfies the conditions a), b), c) on page so that the limit $X(t,\omega)$ given by the expression (12) exists. Now the difference $L(t,\omega) - X(t,\omega)$ is an additive process with continuous sample function and it follows that the difference must be of the form $mt + \sigma B(t)$, where $B(t)$ is a Brownian motion. The crucial point is to show that $X(t)$, $0 \leq t \leq 1$, and $B(t)$, $0 \leq t \leq 1$, are mutually independent. This requires several steps, each rather elementary, which we omit.

3.4. Stable processes

A Lévy process with stationary increments gives a probability measure on the space of ruled functions vanishing at $t = 0$. We are interested in the study of such measures, in particular the characterization of measures from the point of view of stability under a group or a semi-group of transformations acting on function spaces.

In terms of transformation (semi-) groups the property of having stationary increments can be characterized in the following way.

1°) Semigroup of shifts.

Let s_τ, $\tau \geq 0$, be a transformation acting on ruled functions:

$$s_\tau: f(t) \longrightarrow f(t+\tau) - f(\tau) \equiv (s_\tau f)(t) .$$

The collection $\underline{\underline{S}} = \{s_\tau : \tau \geq 0\}$ forms a semigroup under composition

$$s_\tau s_\sigma = s_{\tau+\sigma} , \quad \tau, \sigma \geq 0 . \tag{17}$$

For each sample function of a Lévy process $L(t,\omega)$ we can s_τ to get a new Lévy process $(s_\tau L)(t,\omega)$ identical in law to $L(t$

An additive process with ruled sample functions is a Lévy process with stationary increments if and only if the induced measure on function space is invariant under $\underline{\underline{S}}$.

2°) Group of tensions

Let us consider a transformation acting on the argument of sample functions (i.e. time). Since we always wish to preserve stationary increments, the only reasonable transformation is a tension $g_a : t \longrightarrow at$, $a > 0$. Thus for a given Lévy process $L(t,\omega) \equiv L(at,\omega)$. But there is no Lévy process the distribution of which is invariant under g_a , except the trivial case $L(t,\omega) \equiv 0$. Thus we are led to consider classes of Lévy processes invariant under g_a , $a > 0$.

Now the collection $\underline{\underline{G}} = \{g_a ; a > 0\}$ forms an abelian group with the multiplication

$$g_a g_b = g_{ab} . \tag{18}$$

$\underline{\underline{G}}$ is called the tension group.

Since constant multiples are not important, we shall classify stochastic processes by the following relation: Two stochastic processes $X(t)$, $t \in T$ and $Y(t)$, $t \in T$ are said to be of the same type if $X(t)$, $t \in T$ and $cY(t)$, $t \in T$ have the same distribution with some positive constant c . The class of stochastic processes containing $X(t)$ will be denoted by $\tilde{X}$.

We shall restrict our attention to the classification of Lévy processes with stationary increments and with time parameter space $[0,\infty)$. Note that the shift and the tension are well defined for a class of Lévy processes, i.e. each operation carries a class to another class. We note further that the classification of Lévy processes with stationary increments naturally induces a corresponding classification of ψ-functions. In fact ψ_1 and ψ_2 belong to the same class if and only if $\psi_1(z) = \psi_2(cz)$ for some positive constant c .

We prove

<u>Theorem 3.6.</u> If a class of Lévy processes with stationary increments is invariant under the tension group $\underline{\underline{G}}$, then the corresponding class of ψ-functions is determined by one of the following ψ-functions:

$$(19)\quad \begin{cases} \text{i)} & imz, \quad m \text{ real} \\ \text{ii)} & -\frac{1}{2}z^2 \\ \text{iii)} & \left(-1+i\frac{z}{|z|}a\right)|z|^{\alpha}, \quad a \text{ real}, \quad 0 < \alpha < 2 . \end{cases}$$

<u>Proof.</u> Suppose that a class $\tilde{X}$ of Lévy processes is invariant under $\underline{\underline{G}}$. Let $X(t)$ be any Lévy process in $\tilde{X}$. Then by assumption $X(t)$, $0 \le t < \infty$, and $cX(at)$, $0 \le t < \infty$, have the same distribution with some $c > 0$ (i.e. they are of the same type). The constant c depends on a and we denote it by $1/c(a)$. If ψ is the ψ-function corresponding to $X(t)$, then ψ must satisfy

$$(20)\qquad \psi(z) = a\psi(z/c(a)) .$$

Because of (18) we have

(21) $\psi(z/c(ab)) = \psi(z/c(a)c(b))$ for every z real.

Let us first discuss exceptional cases, namely

Case 1. $\psi(z) \equiv 0$. Then $X(t,\omega) \equiv 0$ a.s.

Case 2. $c(a) \equiv 1$. Then $X(t)$, $0 \leq t > \infty$ and $X(at)$, $0 \leq t < \infty$ have the same distribution. It follows that $X(t,\omega) \equiv 0$ a.s.

Suppose Case 1 and 2 are not true. Then the equation (21) implies that

$$(22) \qquad c(ab) = c(a)c(b) \ .$$

Because, if $\psi(\alpha z) = \psi(z)$ for every z with positive $\alpha < 1$, then $\psi(\alpha^n z) = \psi(z)$, $n = 1,2,\ldots$, and so $0 = \psi(0) \equiv \psi(z)$ which was excluded.

Since $c(a)$, $a > 0$, must be a positive continuous function (see expression (20)) we conclude that $c(a)$ is expressed in the form

$$c(a) = a^{1/\alpha}, \quad \alpha > 0 \ .$$

Thus, by (20), we have

$$\psi(z) = z^{\alpha}\psi(1) \ , \quad z > 0 \ .$$

Noting that $\psi(-z) = \overline{\psi(z)}$, we have

$$\psi(z) = |z|^{\alpha}\overline{\psi(1)} \ , \quad z < 0 \ .$$

Set $\psi(1) = -a_0 + ia_1$, a_0, a_1 real. Then $\psi(z)$ can be expressed in the form

$$\psi(z) = (-a_0 + ia_1 \frac{z}{|z|})|z|^\alpha , \tag{23}$$

where a_0, a_1 are real and $\alpha > 0$.

We shall be able to put restrictions on a_0, a_1 and α in the expression (23). Since $|\exp\{\psi(z)\}| \leqq 1$, a_0 has to be positive. Further as is shown in Feller [1, vol. II, XV. 4], α must satisfy $0 < \alpha \leqq 2$.

Suppose that $a_0 = 0$. In order for $\exp\{ia_1 z \cdot |z|^{\alpha-1}\}$ to be a characteristic function α must equal 1 . Thus $\psi(z) = ia_1 z$, which is the case i) of (19).

Consider another exceptional case: $\alpha = 2$. Then a_1 must be zero and the distribution is Gaussian, which corresponds to ii) of (19).

For the general case the ψ-function in question is equivalent to iii) of (19). Thus the theorem is proved. //

Now we ask whether there exists a Lévy process the ψ-function of which is of the same type as in (19). For the cases i) and ii) the answer is obviously yes. For the case iii), we use the following examples to give an affirmative answer.

Example 1. The case $0 < \alpha < 1$. Consider a Lévy process with Lévy measure n_α given by

$$dn_\alpha(u) = \begin{cases} u^{-\alpha-1}du , & u > 0 \\ 0 , & u < 0 . \end{cases}$$

(See Example 3 in §3.3.)

It, of course, satisfies $\int \frac{u^2}{1+u^2} dn_\alpha(u) < \infty$, and we can form a Lévy process $X_\alpha(t,\omega)$ by the method of Theorem 3.2. Consider a Lévy process $Y_\alpha(t,\omega) = X_\alpha(t,\omega) + t \int \frac{u}{1+u^2} dn_\alpha(u)$ (since $\alpha < 1$, $Y_\alpha(t,\omega)$ is well defined). The ψ-function of $Y_\alpha(t)$ is computed as follows:

$$\psi_\alpha(z) = \int_0^\infty (e^{izu} - 1 - \frac{izu}{1+u^2}) \frac{du}{u^{\alpha+1}} - iz \int_0^\infty \frac{u}{1+u^2} \cdot \frac{du}{u^{\alpha+1}}$$

$$= \int_0^\infty (e^{izu} - 1) \frac{du}{u^{\alpha+1}} = |\Gamma(-\alpha)(-\cos \frac{\pi}{2} \alpha + i \frac{z}{|z|} \sin \frac{\pi}{2} \alpha)|z|^\alpha .$$

If we take the inverted measure n'_α of n_α , i.e.

$$dn'_\alpha(u) = \begin{cases} 0 , & u > 0 \\ |u|^{-\alpha-1}du , & u < 0 \end{cases}$$

we obtain a Lévy process $Y'_\alpha(t,\omega)$ which has a ψ-function

$$\psi'_\alpha(z) = |\Gamma(-\alpha)|(-\cos \frac{\pi}{2} \alpha - i \frac{z}{|z|} \sin \frac{\pi}{2} \alpha)|z|^\alpha .$$

We can always form a Lévy process $Z_\alpha(t,\omega)$ of the form $Z_\alpha(t,\omega) = c_+ Y_\alpha(t,\omega) + c_- Y'_\alpha(t,\omega)$. With a suitable choice of c_+ and c_- this has the ψ-function $(-a_0 + ia_1 \frac{z}{|z|})|z|^\alpha$ for any given $a_0 > 0$

and a_1. This is the same type as the ψ-function given by (19) iii) with $\alpha < 1$.

It is obvious that $Z_\alpha(at,\omega)$, $0 \leq t < \infty$ and $a^{1/\alpha} Z_\alpha(t,\omega)$, $0 \leq t < \infty$, have the same distribution, i.e. $Z_\alpha(at)$ and $Z_\alpha(t)$ are of the same type.

Example 2. The case $1 < \alpha < 2$. Consider a ψ-function

$$\psi(z) = c_+ \int_0^\infty (e^{izu} - 1 - izu) \frac{du}{u^{\alpha+1}} + c_- \int_{-\infty}^0 (e^{izu} - 1 - izu) \frac{du}{|u|^{\alpha+1}} .$$

If we consider

$$\psi'(z) = \psi(z) + iz \left\{ c_+ \int_0^\infty \left(u - \frac{u}{1+u^2}\right) \frac{du}{u^{\alpha+1}} + c_- \int_{-\infty}^0 \left(u - \frac{u}{1+u^2}\right) \frac{du}{|u|^{\alpha+1}} \right\}$$

then the Lévy process $X(t)$ with ψ-function ψ' can be formed by

Theorem 3.2. Then the Lévy process

$$Z(t,\omega) = X(t,\omega) + t \left\{ c_+ \int_0^\infty \frac{u^3}{1+u^2} \frac{du}{u^{\alpha+1}} + c_- \int_{-\infty}^0 \frac{u^3}{1+u^2} \frac{du}{|u|^{\alpha+1}} \right\}$$

has the given function ψ as ψ-function.

Computations similar to those of Example 1 show that $\psi(z)$ can be expressed in the form

$$\psi(z) = (-a_0 + ia_1 \frac{z}{|z|})|z|^\alpha , \quad a_0 > 0 , \quad a_1 \text{ real}$$

and that for any pair (a_0, a_1) we can find a suitable pair (c_+, c_-).

Finally we note that $Z(at,\omega)$, $0 \leq t < \infty$, and $a^{1/\alpha}Z(t,\omega)$, $0 \leq t < \infty$ have the same distribution.

<u>Example 3.</u> The case $\alpha = 1$. Let us begin with the equality

$$-\pi|z| = \int_{-\infty}^{\infty} (e^{izu} - 1 - \frac{izu}{1+u^2}) \frac{du}{u^2} .$$

Again, Theorem 3.2 guarantees the existence of a Lévy process with ψ-function $-a_0|z|$, $a_0 > 0$. Just by adding linear non-random term we can prove that there always exists a Lévy process $Z(t,\omega)$ having the ψ-function of the form $-a_0|z| + ia_1 z$ for any $a_0 > 0$ and a_1 .

The above examples give a complete affirmative answer to the question of existence of a Lévy process for any ψ-function of the form (19).

Note that for any ψ-function given by (19) the corresponding Lévy process $Z(t)$ enjoys the property that

(24) $$Z(at,\omega),\ 0 \leq t < \infty , \quad \text{and} \quad a^{1/\alpha}Z(t,\omega),\ 0 \leq t < \infty$$

have the same distribution. In view of (24), $Z(t)$ is called a <u>stable process</u> and α is called the characteristic <u>exponent</u> of the stable process. Further, if $\psi(z) = \psi(-z)$, i.e. $\psi(z) = -a_0|z|^{\alpha}$, then the stable process is called <u>symmetric</u>. If the Lévy measure is supported on $(0,\infty)$ (e.g. $X_{\alpha}(t,\omega)$ in Example 1) then the process is said to be <u>increasing</u> or (positive) <u>one-sided</u>.

In terms of transformation groups, we can state the following.

Corollary. A minimal class of Lévy processes which is invariant under $\underline{\underline{S}}$ and $\underline{\underline{G}}$ is a class of processes of the same type as a single stable process.

Before leaving this topic, we outline an alternative approach which has more probabilistic content. Our discussion will be divided into three parts.

i) Let $P_\lambda(t,\omega)$ be a Poisson process with parameter $\lambda > 0$, i.e., $E(P_\lambda(t,\omega)) = \lambda t$. Then for $g_a \in \underline{\underline{G}}$ we have

$$g_a : P_\lambda(t,\omega) \longrightarrow (g_a P_\lambda)(t,\omega) = P_\lambda(at,\omega) ,$$

which has the same distribution as $P_{a\lambda}(t,\omega)$. Thus the processes $P_\lambda(t,\omega)$ and $g_a P_\lambda(t,\omega)$ induce measures on function space, the supports of which are mutually disjoint. Also we can see that a system $\textcircled{P_\Lambda} = \{P_\lambda(t,\omega);\ \lambda \in \Lambda\}$ is $\underline{\underline{G}}$ invariant if and only if $\Lambda = (0,\infty)$.

ii) Let the system $\textcircled{P}$ satisfy the conditions a), b), c) on page 38, with the measure n defined on $(0,\infty)$, where the I runs over the set of all intervals in $(0,\infty)$. We require that by $g_a \in \underline{\underline{G}}$ each $P_I(t,\omega)$ should be transformed into a process having the same distribution as some member of $\textcircled{P}$, say $P_{I_a}(t,\omega)$. Since $(g_a P_I)(t,\omega) = P_I(at,\omega)$ has the mean $\tan(I)$, we must have $n(I_a) = an(I)$. Thus it is natural to assume regularity of the mapping

$$(I,a) \longrightarrow I_a \ .$$

In detail, I_a is uniquely determined by the pair (I,a) and if $I = (\xi,\eta]$, then I_a is an interval of the form $(f(a,\xi),g(a,\eta)]$ with smooth functions f and g.

Obviously, by the group property of the g_a, f satisfies

$$(25) \qquad \begin{cases} f(1,\xi) = \xi \\ f(ab,\xi) = f(a,f(b\xi)) \end{cases}$$

and similarly for g.

Lemma. If $f(a,\xi)$ is a smooth function satisfying the relation (25), then f is a function of the form

$$f(a,\xi) = a^{\nu}\xi$$

for some ν.

The proof is easy and is omitted.

The function $g(a,\eta)$ is also expressed in the form $b^{\mu}\eta$ with some μ. But we must have $\nu = \mu$, because $a^{\nu}\xi < a^{\mu}\eta$ whenever $\xi < \eta$. Thus we have

$$I_a = (a^{\nu}\xi,\ a^{\nu}\eta] \quad \text{if} \quad I = (\xi,\eta] \ ,$$

and hence

$$n\{(a^{\nu}\xi,\ a^{\nu}\eta]\} = an\{(\xi,\eta]\} \ .$$

In particular, for $x > 0$, we obtain

$$n\{(x,\infty)\} = x^{1/\nu}n([1,\infty)) = c\cdot x^{1/\nu} \ .$$

Since $n\{(x,\infty)\}$ is decreasing, ν is negative. Set $-1/\nu = \alpha$. Then we have

$$n(I) = \int_I \frac{1}{u^{\alpha+1}} du , \quad \alpha > 0 .$$

By the requirement (10), we restrict α to $0 < \alpha < 2$. We conclude that if (P) is invariant under $\underset{=}{G}$, then the measure n has to be the Lévy measure corresponding to a stable process.

iii) Let (P) be the system determined in ii). Consider the integral based on $P_{du}(t,\omega)$

$$X(t,\omega) = \int_0^\infty (uP_{du}(t,\omega) - tudn(u))$$

the existence of which was proved in §3.3. By $g_a \in \underset{=}{G}$, $uP_{du}(t,\omega)$ is transformed to $uP_{du}(at,\omega)$ which has the same distribution as $uP_{d(a^{-1/\alpha}u)}(t,\omega)$. This, together with the trivial identity $g_a t = at$, gives

$$g_a(uP_{du}(t,\omega) - utdn(u)) \sim uP_{d(a^{-1/\alpha}u)}(t,\omega) - atu \frac{du}{u^{1+\alpha}} ,$$

where $\sim$ means "has the same distribution as". The last expression can be written as $a^{1/\alpha}\left\{ uP_{du}(t,\omega) - tu \frac{du}{u^{1+\alpha}} \right\}$. Thus we have

$$g_a X(t,\omega) \sim a^{1/\alpha} X(t,\omega)$$

which determines the stable process with exponent α.

A similar treatment is possible in the case where n is supported by $(-\infty,0)$ or $(-\infty,0) \cup (0,\infty)$.

3.5 Subordination

We now discuss a transformation which changes the time parameter of sample functions by an increasing stochastic process. Such a <u>random time change</u> already appeared in §3.2[B] although the time parameter is discrete. In fact, if we set $S_n = \sum_{j=0}^{n} \eta_j$ $(S_0 = 0)$, then the compound Poisson process $Y(t,\omega)$ given by (7) is expressible as $S_{P(t,\omega)}(\omega)$, where n in $S_n(\omega)$ is replaced by a Poisson process $P(t,\omega)$. This method is applicable to continuous parameter stochastic processes, in particular to Lévy processes.

Another somewhat direct motivation is the following. Take a Lévy process $X(t,\omega)$ with stationary increments, and form a new process $Y_\xi(t,\omega) = X(\xi(t),\omega)$ by changing the time variable, where $\xi(t)$ is an increasing function with $\xi(0) = 0$. If $\xi(t)$ has a jump, a fixed discontinuity arises for $Y_\xi(t,\omega)$. Therefore, in order to obtain a Lévy process $Y_\xi(t,\omega)$ we must consider a random function $\xi(t)$ such that $\xi(t)$ jumps at any fixed time t_0 with probability 0. Thus we are led to consider an increasing stochastic process $\xi(t,\omega')$, $\omega' \in \Omega'(P')$, and to form $Y(t,\tilde{\omega}) = X(\xi(t,\omega'),\omega)$ where $\tilde{\omega} = (\omega,\omega')$ by extending the probability space from Ω to $\tilde{\Omega} = \{\tilde{\omega}\}$ with probability measure $\tilde{P} = P \times P'$.

Consider a transformation g_ξ acting on a Lévy process.

$$(26) \qquad g_\xi : X(t,\omega) \longrightarrow (g_\xi X)(t,\tilde{\omega}) = X(\xi(t,\omega'),\omega), \qquad \tilde{\omega} \in \tilde{\Omega}$$

where $\xi(t,\omega')$, $0 \le t < \infty$, is an increasing Lévy process with stationary increments. It is easy to show that the process $(g_\xi X)(t,\tilde{\omega})$ is again a Lévy process with stationary increments. Now the collection $\Lambda = \{g_\xi\}$ may be considered as an extension of the transformation group $\underline{\underline{G}}$ introduced in §3.4. If $\xi(t,\omega')$ is a deterministic process, i.e. $\xi(t,\omega') = at$ a.s., then g_ξ turns out to be g_a in $\underline{\underline{G}}$.

We introduce the product in Λ :

$$(27)\qquad \begin{gathered} (g_{\xi_2} g_{\xi_1} X)(t,\tilde{\omega}) = g_{\xi_2}\{X(\xi_1(t,\omega'),\omega)\} = X\{\xi_1(\xi_2(t,\omega''),\omega'),\omega\} \\ \tilde{\omega} = (\omega,\omega',\omega'') \in \Omega \times \Omega' \times \Omega''(P \times P' \times P''), \end{gathered}$$

where $\Omega(P)$, $\Omega'(P')$, $\Omega''(P'')$ are probability spaces on which $X(t)$, $\xi_1(t)$, $\xi_2(t)$ are defined respectively. This definition of the product is compatible with the multiplication in $\underline{\underline{G}}$ (i.e. the formula (18)). Therefore $\underline{\underline{G}}$ can be imbedded in Λ as a subgroup. We wish to find a subclass Ξ of Λ satisfying the following conditions:

i) $\xi(t,\omega)$, $g_\xi \in \Xi$, is an increasing Lévy process with stationary increments.

ii) Ξ forms a continuous semigroup or group imbedded in a certain two-dimensional connected Lie group $\underline{\underline{H}}$, and $\underline{\underline{G}}$ is a proper subgroup of Ξ :

$$(28)\qquad \underline{\underline{G}} \subset \Xi \subset \underline{\underline{H}} .$$

First we note that Ξ cannot be abelian. For, if

$$g_\xi g_a = g_a g_\xi \,, \quad \text{for every } a \,,$$

then it can be shown that $g_\xi \in \underline{\underline{G}}$; i.e. we have $\underline{\underline{G}} = \Xi$ which is excluded. Thus the structure of the two-dimensional non-abelian connected Lie group $\underline{\underline{H}}$ is determined uniquely. Indeed, if the basis is chosen appropriately, the Lie algebra of $\underline{\underline{H}}$ is represented by

$$e_1 = \begin{pmatrix} 1 & 0 \\ 0 & 0 \end{pmatrix}, \qquad e_2 = \begin{pmatrix} 0 & 1 \\ 0 & 0 \end{pmatrix},$$

or equivalently we have a matrix representation

$$\underline{\underline{H}} \cong \left\{ \begin{pmatrix} e^x & y \\ 0 & 1 \end{pmatrix} ; \ x, y \ \text{real} \right\} .$$

Corresponding to e_1 and e_2 $\underline{\underline{H}}$ has two one-dimensional subgroups $\underline{\underline{H}}_1$ and $\underline{\underline{H}}_2$ which are isomorphic to

$$\left\{ \begin{pmatrix} e^x & 0 \\ 0 & 1 \end{pmatrix} \right\} \text{ and } \left\{ \begin{pmatrix} 1 & y \\ 0 & 1 \end{pmatrix} \right\}$$

respectively. $\underline{\underline{H}}_2$ is a normal subgroup of $\underline{\underline{H}}$ while $\underline{\underline{H}}_1$ is not normal and so it is natural to give an isomorphism

$$\underline{\underline{G}} \cong \left\{ \begin{pmatrix} 1 & y \\ 0 & 1 \end{pmatrix} ; \ y \ \text{real} \right\} .$$

Consequently there is an injection from $\Xi/\underline{\underline{G}}$ into $\underline{\underline{H}}_1$. Therefore we are given the following commutation relation

$$g_\xi g_a = g_{ae^x} g_\xi , \qquad g_\xi \in \Xi \tag{29}$$

where x is determined uniquely by the coset of $\Xi/\underline{\underline{G}}$ including ξ. x is zero if and only if $g_\xi \in \underline{\underline{G}}$ ($\xi(t)$ is deterministic).

In terms of stochastic processes, (29) can be expressed in the form

$$a\xi(t,\omega) \sim \xi(ae^x t,\omega) .$$

Thus $\xi(t,\omega)$ must be a stable process. By the requirement i) we see that $\xi(t,\omega)$ is a one-sided stable process with exponent $\alpha<1$.

We now determine Ξ explicitly. Since Ξ is two-dimensiona we can take all the stable processes $\xi(t,\omega)$ satisfying

$$\log E(e^{-\lambda\xi(1)}) = -a\lambda^\alpha, \quad a > 0, \quad 0 < \alpha \leqq 1, \quad \lambda > 0 . \tag{30}$$

Obviously Ξ forms a two-dimensional semigroup with the product (27) and $\underline{\underline{G}}$ is the subgroup consisting of all the ξ with $\alpha = 1$ in the expression (30).

Given a Lévy proces $X(t,\omega)$ by the formula (12). By $g_\xi \in \Xi$ we obtain a new Lévy process $(g_\xi X)(t,\tilde{\omega})$ (see (26)). If real $\psi(z)$ is the ψ-function of $X(t)$, the characteristic function of $(g_\xi X)(t)$ is given by

$$\begin{aligned} E[\exp\{izX(\xi(t,\omega'),\omega)\}] &= \int \exp\{\xi(t,\omega')\psi(z)\}dp'(\omega') \\ &= \exp\{-ta|\psi(z)|^\alpha\} . \end{aligned}$$

Thus we have arrived at the Bochner's theory of subordination

(S. Bochner [8, Chapter 14]). The following examples will show how the transformation semigroup Ξ operates.

Example 1. Let $B(t,\omega)$ be a Brownian motion. g_ξ transforms $B(t)$ into a Lévy process with ψ-function $a|z|^{2\alpha}$, where ξ satisfies (30). Hence from a Brownian motion we can form all the symmetric stable processes by Ξ.

Example 2. The collection of the symmetric stable processes with characteristic exponents less than $\alpha_0 (\leqq 2)$ is invariant under Ξ.

Remark. We can also discuss the transformations of a class of Lévy processes. Many interesting topics related to subordination are referred to in W. Feller [I. vol.II].

Bibliography

[8] Bochner, S., Harmonic analysis and the theory of probability. Univ. of California Press, 1955.

[9] Ito, K., Lectures on stochastic processes. Tata Inst. Bombay, 1961.

[10] -----, On stochastic processes (I), Japanese J. Math. 18, (1942), 261-301.

[11] Lévy, P., Fonction aléatoires à corrélation linéaires. Illinois J. Math., vol.1(1957), 217-258.

Part II

§4. Stationary processes

This article starts out by discussing probability measures on the space of (generalized) functions, as well as the need for such measures. This leads to our defining a stationary process in the generalized sense, as a probability measure μ on function space which is invariant under any shift of the argument of the functions. For any such measure μ we have the Hilbert space $L^2(\mu)$ of functionals of sample functions of the given stationary process.

We do some analysis on the space $L^2(\mu)$ in section 4.4. Finally we consider the very important class of stationary process with independent value at every moment which are analogous to sequences of independent identically distributed r.v.'s in the case of discrete time parameter.

4.1. Measures on function spaces

We are interested in analyzing functionals of Lévy processes (see §0). It seems reasonable to discuss functionals not of a Lévy process itself but of the derivative of it, since this looks like a (continuous) sum of independent (infinitesimal) r.v.'s. Unfortunat sample functions of the derivative of a Lévy process are not function in the ordinary sense. This leads us to introduce a probability measure on the space of generalized functions, and thus to generalize the notion of a stochastic process.

There are many ways to approach the problem of introducing a measure on an infinite dimensional vector space and in particular on a function space. In our approach, Bochner's theorem is the most powerful tool. We begin by stating Bochner's theorem in the finite dimensional case.

Theorem 4.1. (S. Bochner) For any probability measure dF (probability distribution) on R^n, the characteristic function

(1) $$\varphi(z) = \int_{R^n} \exp[i(z,x)]dF(x) , \qquad z \in R^n(\cong(R^n)^*)$$

has the following properties:

i) $\varphi(z)$ is continuous,

ii) $\varphi(z)$ is positive definite,

iii) $\varphi(0) = 1$.

Conversely, if a given function $\varphi(z)$, $z \in R^n$, satisfies the conditions i), ii) and iii), then there exists the unique probability measure dF on R^n such that the relation (1) holds.

We are now going to generalize the theorem to the case where the underlying space is a function space which is much bigger than $L^2(R^1,\text{Leb})$ space.

We consider a triple

(2) $$E \subset L^2(T, \text{Leb}) \equiv H \subset E^* , \quad T \text{ interval} \subset R^1 ,$$

that is, a nice topological vector space E such that the dual space E^* includes H. Then the first problem is to find a condition under which a measure μ on E^* can be given by a positive definite continuous function on E which is related to μ by a formula similar to (1). We note that Bochner's theorem does not generally hold in the infinite dimensional case, as the following example shows.

Example 1. Let $E = H$ (hence $E^* = H$) and $C(\xi) = \exp\{-\|\xi\|^2\}$. Obviously C is continuous in H, positive definite and $C(0) = 1$. However, no measure μ on H corresponds to $C(\xi)$ satisfying the relation

$$\exp\{-\|\xi\|^2\} = \int_H \exp\{i\langle x,\xi\rangle\}d\mu(x) .$$

Indeed, suppose a measure μ did exist such that the above relation holds. Then the collection $\{(x,\xi):\ \xi \in H\}$ forms a Gaussian system and so, for any complete orthonormal system $\{\xi_n\}$ in H, $\langle x,\xi_n\rangle$, $n=1,2,\ldots,$ is a sequence of independent Gaussian r.v.'s. Thus the strong law of large numbers implies

$$\lim_{N\to\infty} \frac{1}{N} \sum_{n=1}^{N} \langle x,\xi_n\rangle^2 = 1 , \quad \text{a.s.}(\mu) .$$

(Note that $\langle x,\xi_n\rangle$ has the distribution $N(0,1)$). On the other hand, since $x \in H$,

$$x = \sum_n \langle x,\xi_n\rangle\xi_n$$

and $\|x\|^2 = \Sigma_n \langle x,\xi_n\rangle^2 < \infty$, which leads to a contradiction. This suggests that, if a measure μ did exist, it could not be supported by H. Therefore to catch the support of μ, we extend H, that is, we work with the triple (2) with the measure μ defined on E^*.

First we have to establish a measurable space. Let $\xi_1,\ldots,\xi_n$ be linearly independent in E. For any $x \in E^*$, there is a mapping

$$x \longrightarrow (\langle x,\xi_1\rangle,\ldots,\langle x,\xi_n\rangle) \in R^n ,$$

where $\langle x,\xi\rangle$, $x \in E^*$, $\xi \in E$, denotes the bilinear form which pa[irs] E and E^*. A subset of E^*

(3) $$A_{\xi_1,\ldots,\xi_n,B} = \{x \in E^* ;\ (\langle x,\xi_1\rangle,\ldots,\langle x,\xi_n\rangle) \in B\} ,$$

with an n-dimensional Borel set B is called a <u>cylinder set</u>. If F is a finite dimensional subspace of E and if the ξ_j, $1 \leq j \leq n$, are chosen from F, then a cylinder set of the form (3) is said to b[e] based on F . The collection

$$\text{Ⓐ} = \{A_{\xi_1,\ldots,\xi_n, B} ; n = 1,2,\ldots,\xi_1,\ldots,\xi_n \in E, \quad B \in \text{Ⓑ}(R^n)\}$$

forms a field. The space E^* together with the σ-field Ⓑ generated by Ⓐ is a measurable space on which we are going to introduce our measures.

Given a probability measure μ on $(E^*, \text{Ⓑ})$, the functional $\exp\{i\langle x,\xi\rangle\}$ of x is integrable and the integral

$$C(\xi) = \int_{E^*} \exp\{i\langle x,\xi\rangle\} d\mu(x) \tag{4}$$

gives a function $C(\xi)$ on E which has the following properties:

$$\text{(5)} \qquad \begin{array}{ll} \text{i)} & C(\xi) \text{ is continuous on } E , \\ \text{ii)} & C(\xi) \text{ is positive definite} , \\ \text{iii)} & C(0) = 1 . \end{array}$$

Thus, to establish a generalization of Bochner's theorem it suffices to prove the converse of the above assertion. Let $C(\xi)$, $\xi \in E$, be a functional satisfying the conditions i), ii), iii) of (5). Then our approach may be divided into two parts as follows.

a) We form a finitely additive measure μ on $(E^*, \text{Ⓐ})$ from the given $C(\xi)$ so that the relation (4) holds.

b) We extend μ to a σ-additive measure on $(E^*, \text{Ⓑ})$.

We do this in detail in the next section.

4.2. Extension theorems

a) Let F be an n-dimensional subspace of E spanned by $\xi_1,\ldots,\xi_n$. Denote by F^a the linear subspace of E^* given by

$$F^a = \{x; \langle x,\xi\rangle = 0 \quad \text{for all} \quad \xi \in F\} \quad \text{(annihilator)} .$$

The factor space E^*/F^a is obviously an n-dimensional space. Let ρ_F be the canonical projection

$$\rho_F: E^* \longrightarrow E^*/F^a .$$

A subset A of E^* is a cylinder set based on F if and only if it is expressed in the form $A = \rho_F^{-1}(B)$ with B a Borel subset of E^*/F^a

For the given functional $C(\xi)$ satisfying (5) we consider the restriction $C_F(\xi)$ of $C(\xi)$ to the subspace F. Since $C_F(\xi)$ is continuous on F and is positive definite, Bochner's theorem is appli cable and there exists a probability measure m_F on E^*/F^a ($\cong F^*$) such that

$$C_F(\xi) = \int_{E^*/F^a} \exp\{i\langle\bar{x}, \xi\rangle_F\} dm_F(\bar{x}) , \qquad \xi \in F$$

where the bilinear form $\langle\bar{x}, \xi\rangle_F$ is naturally induced by the original bilinear form $\langle x, \xi\rangle$, $x \in E^*$, $\xi \in F$.

If F and G are two finite dimensional subspaces of E with $F \subset G$, there is a projection T :

$$T: E^*/G^a \longrightarrow E^*/F^a .$$

Since $C_F(\xi)$ is nothing but the restriction of $C_G(\xi)$ to F, we obtain

$$m_F(B) = m_G(T^{-1}B) , \qquad B \subset E^*/F^a .$$

Thus we are given a <u>consistent</u> family $\{m_F$: F is a finite dimensional subspace of $E\}$ of probability measures.

For a cylinder set A we define

$$\mu(A) = m_F(\rho_F(A))$$

if A is based on F. This definition makes sense because of the consistency of the family $\{m_F\}$. Further it is easy to see that the set function μ is well defined on Ⓐ and finitely additive. Noting that $\mu(E^*) = 1$, we have obtained a finitely additive probability measure μ on $(E^*, Ⓐ)$.

b) We now proceed to the second step. First we prepare some lemmas; the first two of them are well known but the third one is fundamental (see R. A. Minlos [12] and I. M. Gelfand-N. Ya. Vilenkin [13]).

Lemma 1. Let Ⓐ be a field of sets and Ⓑ the σ-field generated by Ⓐ. A finitely additive measure μ defined on Ⓐ can be extended to a (σ-additive) measure $\bar{\mu}$ on Ⓑ if and only if μ is σ-additive on Ⓐ.

With the same notations as above, we state

Lemma 2. Let μ be finite. Then μ is completely additive on Ⓐ if and only if, for any decreasing sequence $A_n \in$ Ⓐ and $\bigcap_n A_n = \emptyset$, we have

$$\lim_{n \to \infty} \mu(A_n) = 0 .$$

In this case the extension $\bar{\mu}$ exists and is unique.

Lemma 3. Let μ be a probability measure on $(R^n)^*$ $(\cong R^n)$ with the characteristic function $\varphi(z)$, $z \in R^n$ If $|\varphi(z) - 1| < \varepsilon$ on the set $\mathcal{E} = \{z; \sum_{i=1}^n a_i^2 z_i^2 \leqq \gamma^2\}$, then we have

$$(6) \qquad \mu(S^c) < \beta^2(\varepsilon \frac{2}{\gamma^2 t^2} \sum_{i=1}^n a_i^2)$$

where S is a ball with radius t and β is a universal constant.

Proof. We have

$$\int_{R^n} [1-\exp(-\frac{1}{2t^2} \sum_1^n x_i^2)] d\mu(x) \geqq \int_{S^c} [1-\exp(-\frac{1}{2t^2} \sum_1^n x_i^2)] d\mu(x)$$
$$\geq (1-e^{-\frac{1}{2}})\mu(S^c) .$$

On the other hand,

$$\int_{R^n} [1-\exp(-\frac{1}{2t^2} \sum_1^n x_i^2)] d\mu(x) = 1 - \int_{R^n} \exp(-\frac{1}{2t^2} \sum_1^n x_i^2) d\mu(x)$$

$$= 1 - (\frac{t^2}{2\pi})^{n/2} \int_{R^n} \varphi(z) \exp(-\frac{t^2}{2} \sum_1^n z_i^2) dz$$

$$= (\frac{t^2}{2\pi})^{n/2} \int_{R^n} (1-\varphi(z)) \exp(-\frac{t^2}{2} \sum_1^n z_i^2) dz$$

$$\leqq (\frac{t^2}{2\pi})^{n/2} \left| \int_{\mathcal{E}} " + \int_{\mathcal{E}^c} \right|$$

$$< \varepsilon + (\frac{t^2}{2\pi})^{n/2} \frac{2}{\gamma^2} \int_{\mathcal{E}^c} (\sum_1^n a_i^2 z_i^2) \exp(-\frac{t^2}{2} \sum_1^n z_i^2) dz$$

$$< \varepsilon + \frac{1}{\gamma^2 t^2} \sum_1^n a_i^2 .$$

Thus we have

$$\mu(S^c) < \frac{1}{1-\sqrt{e}^{-1}} (\varepsilon + \frac{1}{\gamma^2 t^2} \sum_1^n a_i^2) .$$

We return to the original problem. Assume that the topology of the vector space E in the expression (2) is given by the consistent family of Hilbertian norms $\| \|_n$, and that E is complete with respect to this topology. We denote by E_n the Hilbert space obtained by the completion of E with respect to the norm $\| \|_n$. Then, by assumption, we have

$$E = \bigcap_n E_n .$$

Such a space E is called a <u>σ-Hilbert space</u>. We may assume, without loss of generality, that the norms $\| \|_n$ are increasing:

$$\| \|_1 < \| \|_2 < \ldots < \| \|_n < \ldots .$$

This implies that

$$H \supset E_1 \supset E_2 \supset \ldots \supset E_n \supset \ldots ,$$

and hence

$$H = H^* \subset E_1^* \subset E_2^* \subset \ldots \subset E_n^* \subset \ldots .$$

We denote by $\| \|_{-n}$ the norm in E_n^*.

Hereafter E is always assumed to be σ-Hilbert and the norms $\| \|_n$, $-\infty < n < \infty$, defining the topology of E and E^* to be increasing.

<u>Lemma 4.</u> In order that a finitely additive measure μ on $(E^*, Ⓐ)$ have an extension $\bar{\mu}$ on $(E^*, Ⓑ)$ which is σ-additive, it is necessary and sufficient that for any $\varepsilon > 0$ there exists a ball $S_n = \{x; \|x\|_{-n} \leqq \gamma_n\}$ in E^* for some n such that for any cylinder set A outside of S_n we have

$$\mu(A) < \varepsilon .$$

<u>Proof.</u> Suppose μ has an extension $\bar{\mu}$. Choose the balls S_n with $\gamma_n \longrightarrow \infty$. Then we have

$$\bigcup_n S_n = E^* ,$$

which shows that $\bar{\mu}(S_n^c) \longrightarrow 0$. Hence, for any $\varepsilon > 0$, we can find S_n such that $A \cap S_n = \emptyset$ implies that $\mu(A) < \varepsilon$.

Next we prove the converse. Let A_n be mutually disjoint cylinder sets with $\bigcup_n A_n = E^*$. Then, of course,

$$\sum_n \mu(A_n) \leqq 1 .$$

Suppose "=" fails to hold in the above expression, say

$$\sum_n \mu(A_n) = 1 - 3\varepsilon < 1 .$$

For each A_n, we can find open cylinder set A_n' such that $A_n' \supset A_n$ and

$$\mu(A_n' - A_n) < \frac{\varepsilon}{2^n} .$$

Since S_n is (weakly) compact, and since $\bigcup_n A_n' \supset S_n$, we can choose a finite number of the A_n', say $A_1', \ldots, A_k'$, whose union covers S_n.

Set $A' = \bigcup_{n=1}^{k} A'_n$. Then we have

$$1 = \mu(A' + A'^c) = \mu(A') + \mu(A'^c) ,$$

$$\mu(A') \leqq \sum_{n=1}^{k} \mu(A_n) + \varepsilon .$$

By assumption

$$\mu(A'^c) < \varepsilon ,$$

and so

$$1 \leqq \sum_{n=1}^{k} \mu(A_n) + \varepsilon + \varepsilon \leqq (1 - 3\varepsilon) + 2\varepsilon = 1 - \varepsilon$$

which is a contradiction.

Theorem 4.2. Let $C(\xi)$ be a functional on E satisfying

i) $C(\xi)$ is continuous with respect to the norm $\| \|_m$,

ii) $C(\xi)$ is positive definite,

iii) $C(0) = 1$.

If there exists $n(> m)$ such that the injection $T^n_m: E_n \longrightarrow E_m$ is Hilbert-Schmidt, then there is a σ-additive measure $\bar{\mu}$ which is the unique extension of μ satisfying (4) and is supported by E^*_n .

Remark on Hilbert-Schmidt operators. Let H_1 and H_2 be Hilbert sp[aces] and $A: H_1 \longrightarrow H_2$ be completely continuous operator. Then A has a polar decomposition $A = UT$, where $T: H_1 \longrightarrow H_2$ is completely continuous, symmetric and positive, and U is an isometry $TH_1 \longrightarrow H_2$. The operator T has a spectral decomposition of the form

$$Tf = \sum \lambda_k (f, e_k) e_k , \quad \lambda_k \geq 0,$$

where $\{e_k\}$ is the C.O.N.S. in H_1 . Then A is Hilbert-Schmidt if $\|A\|_2^2 = \sum_k \lambda_k^2 < \infty$. $\|A\|_2$ is called the Hilbert-Schmidt norm. If a stronger condition $\sum \lambda_k < \infty$ is satisfied, A is called nuclear.

<u>Proof of Theorem 4.2.</u> For any ε, by assumption, there is a ball (neighborhood of 0) U with radius γ in E_m such that

(7) $$|C(\xi) - 1| < \frac{\varepsilon}{2\beta^2} \text{ for every } \xi \in U$$

(β is the constant appearing in the inequality (6) in Lemma 3.) Then, by hypothesis, there exists a neighborhood V of 0 in E_n such that $T_m^n V \subset U$.

Now take a ball S_n in E_n^* with radius $t = \dfrac{2\beta}{\sqrt{\varepsilon\gamma}} \|T_m^n\|_2$.

S_n is the desired ball satisfying the condition in Lemma 4. To show this, let A be a cylinder set based on a finite dimensional subspace F of E such that $A \cap S_n = \emptyset$. Since A is expressed in the form $A = \rho_F^{-1}(B)$, the last equality implies

(8) $$B \cap \rho_F(S_n) = \emptyset .$$

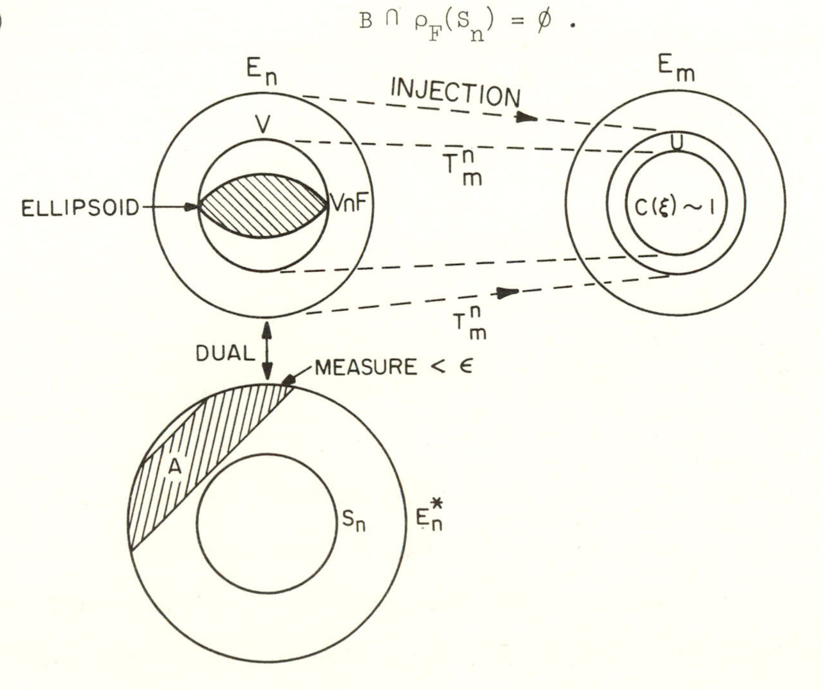

The functional $C_F(\xi)$, the restriction of $C(\xi)$ to F, is a characteristic function and determines a probability distribution m_F on E^*/F^a. C_F still satisfies the inequality (7) for every $\xi \in V \cap F$. Since T_m^n is Hilbert-Schmidt, $V \cap F$ is a (finite dimensional) ellipsoid in the $n^{\underline{th}}$ norm. Therefore, with a suitable choice of C.O.N.S., each $\xi \in V \cap F$ has the coordinate representation: $\xi = (z_1, z_2, \dots, z_k)$, and the coordinates satisfy $\sum_{i=1}^{k} a_i^2 z_i^2 \leqq \gamma^2$. (Recall that $T_m^n V \subset U$), where $\sum a_i^2 \leqq \|T_m^n\|^2$.

We are now ready to apply Lemma 3. By condition (8), we have

$$m_F(\rho_F(S_n)^c) < \beta^2\left(\frac{\varepsilon}{2\beta^2} + \frac{2}{\gamma^2 t^2} \sum a_i^2\right) < \frac{\varepsilon}{2} + \frac{2\beta^2}{\gamma^2 t^2} \|T_m^n\|_2^2 = \varepsilon ,$$

or equivalently

$$\mu(A) = m_F(B) \leqq m_F(\rho_F(S_n)^c) < \varepsilon .$$

The uniqueness is obvious $(\mu(E_n^*) = 1)$ and the theorem is proved.

Let E be a σ-Hilbert space, with topology generated by an increasing sequence of norms $\| \|_n$. If for every m, there exists n such that the injection $T_m^n\colon E_n \longrightarrow E_m$ is Hilbert-Schmidt, then E is called <u>nuclear.</u>

<u>Examples of a nuclear space</u>

i) $\mathcal{S} = \{\xi;\ \xi \in C^\infty,\ \xi(t)$ is rapidly decreasing as $t \longrightarrow \pm\infty\}$

ii) $\mathcal{D} = \{\xi;\ \xi \in C^\infty,\ \xi$ has compact support$\}$

iii) $\mathcal{D}[K] = \{\xi;\ \xi \in C^\infty$, the support of $\xi \subset K\}$, K compact.

For topologies of these spaces see [13].

iv) $C^\infty[K] = \{\xi;\ \xi \in C^\infty(G)$ for some open $G \supset K\}$, K compact.

The topology of $C^\infty[K]$ is given by the following norms (see [13,Ch

$$\|\xi\|_p = \max_{k \leqq p} \max_{t \in K} |\xi^{(k)}(t)| .$$

v) $C^\infty(R^1)$. This space is an inductive limit of the $C^\infty[K_a]$, $K_a = [-a, a]$.

The following theorem can also be proved (see [12] and [13]).

Theorem 4.3. Let E be a nuclear space. Given a continuous, positive definite functional $C(\xi)$ with $C(0) = 1$, then there exists a unique probability measure μ on $(E^*, \textcircled{B})$ such that

$$C(\xi) = \int_{E^*} \exp[i < x, \xi >] d\mu(x) .$$

Definition. Let $C(\xi)$ be as above. Then the measure space (E^*, μ) is called a generalized stochastic process or simply a stochastic process if no confusion arises. For a stochastic process we call each member x of E^* a sample function, and call $C(\xi)$ the characteristic functional of (E^*, μ) .

Remark. In a similar manner we can establish the Kolmogorov extension theorem.

4.3. Definition of a stationary process

Any ξ in E has a coordinate representation: $\xi = (\xi(t), -\infty < t < \infty)$. Define a transformation S_t , $-\infty < t < \infty$, acting on E by

$$S_t: \xi(.) \longrightarrow \xi(\cdot - t) .$$

From now on the nuclear space E will be supposed to satisfy the following

Assumptions

i) E is S_t-invariant,

ii) each S_t is a linear isomorphism of E ,

iii) $< S_t\xi, x >$ is continuous in t for every ξ and x,

iv) E contains functions with compact support.

By assumption ii), a transformation T_t acting on E^* is determined by S_t as follows:

$$< x, S_t\xi > = < T_t x, \xi > .$$

$\{T_t, -\infty < t < \infty\}$ satisfies

$$T_t T_s = T_{t+s} ,$$

that is $\{T_t\}$ forms a one-parameter group. Obviously every T_t leav the field Ⓐ of cylinder sets invariant.

From assumption iii) it follows easily that $f(x,t) \equiv T_t x$ is (x,t) - (Ⓑ $\times$ Ⓑ (R')-) measurable.

<u>Definition.</u> If the measure μ is T_t-invariant, that is $\mu \circ T_t = \mu$ fo every t, then $\mathbb{P} = (E^*, \mu, \{T_t\})$ is called a <u>stationary process.</u> Th characteristic functional of (E^*, μ) is said to be the characteristic functional of $\mathbb{P}$.

The measure μ is T_t-invariant if and only if the characteristi functional is S_t-invariant.

<u>Examples of a stationary process.</u> Let us take $\mathcal{D}$ or $\mathcal{S}$ for the space E so that our assumptions are satisfied.

i) If $C(\xi) = \exp\{-\frac{1}{\alpha}\int|\xi(t)|^{\alpha}dt\}$, $0 < \alpha < 2$, then $C(\xi)$ is S_t-invariant and therefore determines a stationary process which is called a <u>stable white noise</u> (see §11.3.)

ii) Given $C(\xi) = \exp\{\iint (e^{i\xi(t)u} - 1 - \frac{i\xi(t)u}{1+u^2})dn(u)dt\}$, with Lévy measure $dn(u)$ (see §3.3, p40), then we have a statio process which, roughly speaking, is the derivative of a Lév process with Lévy measure $dn(u)$.

iii) Let $X(t,\omega)$, $-\infty < t < "$, $\omega \in \Omega(P)$, be an ordinary strictly stationary process whose sample functions are locally summable. Then $\int X(t,\omega)\, \xi(t)dt$ is a continuous linear functional of $\xi \in \mathcal{D}$. Hence a functional defined by

$$C(\xi) = \int \exp(i \int X(t,\omega)\xi(t)dt)dP(\omega)$$

becomes a characteristic functional of a stationary process $\mathbb{P} = (\mathcal{D}^*,\mu,\{T_t\})$. Thus the process $X(t,\omega)$ can be regarded as a stationary process in our sense.

Let $\mathbb{P} = (E^*,\mu,\{T_t\})$ be a stationary process. The transformation group $\{T_t: -\infty < t < \infty\}$ is a <u>flow</u> on the measure space $(E^*, \textcircled{B}, \mu)$ and therefore it determines a one-parameter group of unitary operators $\{U_t; -\infty < t < \infty\}$ acting on $(L^2) = L^2(E^*,\mu)$. In fact, U_t is defined by

$$(U_t\varphi)(x) = \varphi(T_t x) , \quad \varphi \in (L^2) \tag{9}$$

and $\{U_t\}$ satisfies

$$U_t U_s = U_{t+s} .$$

Moreover, since $T_t x$ is (x,t)-measurable, the operators U_t are strongly continuous. Thus to any stationary process there is always associated a continuous one-parameter group of unitary operators.

4.4. Hilbert spaces (L^2) and $\mathcal{F}$ arising from a stationary process.

Let $P = (E^*, \mu, \{T_t\})$ be a stationary process. Then, as shown above, there is associated with it a Hilbert space $(L^2) = L^2(E^*, \mu)$ and a continuous one-parameter group $\{U_t,\ t \text{ real}\}$ of unitary operators.

Lemma 5. Let A be the algebra spanned by the $e^{i\langle x, \xi \rangle}, \xi \in E$. Then A is dense in (L^2).

Proof. It suffices to show that a function $\varphi(x) \in (L^2)$ orthogonal to every function of the form $\prod_{k=1}^{n} e^{it_k \langle x, \xi_k \rangle}$ is the zero element of (L^2).

Suppose

$$(10) \quad \int_{E^*} e^{i \sum_1^n \langle x, \xi_k \rangle t_k} \overline{\varphi(x)}\, \mu(dx) = 0, \text{ for every } t_k, 1 \leqq k \leqq n.$$

Let $Ⓑ_n$ be the σ-field $Ⓑ(\langle x, \xi_1 \rangle, \ldots, \langle x, \xi_n \rangle)$. The assumption (10) implies that

$$\int e^{i \sum \langle x, \xi_k \rangle t_k} \overline{E(\varphi(x)/Ⓑ_n)}\, \mu(dx) = 0, \text{ for every } t_k,$$

which means Why?

$$E(\overline{\varphi(x)}/ \textcircled{B_n}) = 0, \quad \text{a.e.}$$

Letting $\textcircled{B_n} \uparrow \textcircled{B}$, we obtain

$$E(\overline{\varphi(x)}/\textcircled{B}) = 0.$$

Since $\varphi(x)$ is measurable with respect to $\textcircled{B}$, we conclude that $\varphi(x) = 0$, a.s.

Obviously A is U_t-invariant:

$$U_t A = A \quad \text{for every} \quad t.$$

Next we study the function

$$(11) \qquad C(\xi) = \int e^{i <x,\xi>} d\mu(x)$$

which is analogous to the usual Fourier-Stieltjes transform of a measure on Euclidean space. We define the transformation τ by

$$(12) \quad (\tau\varphi)(\xi) = \int_{E^*} e^{i<x,\xi>} \varphi(x)d\mu(x) \ , \ \varphi \in (L^2) \ , \ \xi \in E.$$

The collection $\mathcal{F} = \{\tau\varphi;\ \varphi \in (L^2)\}$ is a vector space, each member of which is a functional defined on E. We note that

$$(13) \qquad \tau(e^{i\langle x,\eta\rangle})(\cdot) = C(\cdot + \eta).$$

Lemma 6. The transformation τ is an isometry relative to the inner product on $\mathcal{F}$ determined by

$$(14) \qquad (f(\cdot),\ C(\cdot - \xi)) = f(\xi),\ f \in \mathcal{F}.$$

The proof is obvious because of (13) and Lemma 5.

Given a positive definite function $\Gamma(s,t)$, s, $t \in F$ (F may be any abstract space), we can form a Hilbert space $\mathcal{F} = \mathcal{F}(F,\Gamma)$ satisfying the following conditions.

i) $\Gamma(\cdot,t)$ belongs to $\mathcal{F}$ for every t,

ii) $\langle f(\cdot), \Gamma(\cdot,t)\rangle = f(t)$, $f \in \mathcal{F}$, where $\langle\ ,\ \rangle$ is the inner product introduced in $\mathcal{F}$,

iii) $\mathcal{F}$ is spanned by $\{\Gamma(\cdot,t),\ t \in F\}$.

The Hilbert space $\mathcal{F}$ is called a reproducing kernel Hilbert space, and Γ is called the reproducing kernel of $\mathcal{F}$. (See N. Aronszajn [15].)

The last lemma leads us to consider the reproducing kernel Hilbert space $\mathcal{F} = \mathcal{F}(E,C)$. Thus we obtain

Theorem 4.4. The transformation τ defined by (12) is a unitary transformation of (L^2) onto the reproducing kernel Hilbert space $\mathcal{F}(E,C)$.

Corresponding to the unitary operator U_t on (L^2) we introduce an operator

$$\tilde{U}_t = \tau U_t \tau^{-1}.$$

It is quite easy to see that $\{\tilde{U}_t, t \text{ real}\}$ is a continuous one-parameter group of unitary operators on $\mathcal{F}$.

We now come to the analysis on Hilbert spaces (L^2) and $\mathcal{F}$. First of all we shall introduce some basic and elementary concepts such as polynomials, exponential functions and differential operators.

1) Polynomials. Let $P(t_1,t_2,\ldots,t_n)$ be a complex coefficient polynomial on R^n, and let $\xi_1,\xi_2,\ldots,\xi_n$ be in E. A function $\varphi(x) = P(\langle x, \xi_1\rangle, \langle x, \xi_2\rangle, \ldots, \langle x,\xi_n\rangle)$ is a polynomial on E^*. We can speak of the degree of the polynomial φ assuming that ξ_i's are linearly independent, i.e.

$$\text{degree of } \varphi = \text{degree of } P.$$

The collection of all the polynomials is denoted by M.

We further assume the following

Assumptions

v) $$\int |<x,\xi>|^p d\mu(x) < \infty, \; p = 1,2,\ldots,$$

vi) $$\int <x,\xi> d\mu(x) = 0.$$

With the assumption v) we have

$$M \subset (L^2)$$

or, equivalently, we have that $\tau(M)$ is defined and

$$\tau(M) \subset \mathcal{F} .$$

2) Exponential functions. The algebra A spanned by exponential functions $e^{i<x,\xi>}$, $\xi \in E$, was discussed in Lemma 5.

3) Functional derivatives. Consider a multiplication

$$<x,\xi> \cdot \varphi(x) \quad , \quad \varphi \in (L^2).$$

Lemma 7. If $\varphi(X)$ and $<x,\xi> \varphi(x)$ belong to (L^2), then we obtain

(15) $$\tau(<x,\xi> \varphi(x))(\eta) = \frac{1}{i} (D_\xi f)(\eta),$$

where $f = \tau\varphi$ and D_ξ stands for the dunctional derivative defined by

$$(D_\xi f)(\eta) = \lim_{\varepsilon \to 0} \frac{1}{\varepsilon} \{f(\eta + \varepsilon \xi) - (\eta)\}.$$

The domain of the operator D_ξ is denoted by $\mathcal{D}(D_\xi)$. We define $\mathcal{D}$ as $\bigcap_{\xi \in E} \mathcal{D}(D_\xi)$

Theorem 4.5. If the Assumption v) is satisfied, we have the following:

i) $C(\cdot-\xi) \in \mathcal{D}$ for every $\xi \in E$, and $\prod_{j=1}^{n} D_{\xi_j} C(\cdot-\xi)$ belongs to $\mathcal{D}$ for any n and $\xi_1,\dots,\xi_n \in E$,

ii) $\tau(M) \subset \mathcal{D}$,

iii) for any $\xi_1,\dots,\xi_n \in E$ and any choice of positive integers $k_1,\dots,k_n$, we have

$$\tau^{-1}\{\prod_{j=1}^{n} D_{\xi_j}^{k_j} C(\cdot)\} = i^{-\Sigma k_j} \prod_{j=1}^{n} <x, \xi_j>^{k_j},$$

iv) the operator $\frac{1}{i} D_\xi$ is self-adjoint,

v) for any $f \in \tau(M)$ and $\xi_1, \xi_2 \in E$, we have

$$D_{\xi_1} D_{\xi_2} f = D_{\xi_2} D_{\xi_1} f,$$

vi) if $\prod_{j=1}^{n} <x, \xi_j>^{k_j} \varphi(x) \in (L^2)$, we have

$$\tau\left(\prod_{j=1}^{n} \langle x, \xi_j \rangle^{k_j} \varphi(x)\right) = i^{-\Sigma k_j} \prod_{j=1}^{n} D_{\xi_j}^{k_j} f$$

with $\tau\varphi = f$.

Proof. These assertions can be proved by simple computations.

For related topics we refer to Hida-Ikeda [14].

4.5. Stationary processes with independent values at every moment.

We shall discuss a particular class of stationary processes which is a continuous analogue of a sequence of independent r.v.'s.

Definition. A stationary process $P = (E^*, \mu, \{T_t\})$ is said to have independent values at every moment if the characteristic functional C of P satisfies

(16) $C(\xi_1+\xi_2) = C(\xi_1)\cdot C(\xi_2)$ whenever $\mathrm{supp}(\xi_1) \cap \mathrm{supp}(\xi_2) = \emptyset$.
($\mathrm{supp}(\xi)$ stands for the support of ξ).

The property (16) is equivalent to

(16') $\langle x, \xi_1 \rangle$ and $\langle x, \xi_2 \rangle$ are mutually independent r.v.'s on (E^*, μ) if $\mathrm{supp}(\xi_1) \cap \mathrm{supp}(\xi_2) = \emptyset$.

Example. A stationary process with a characteristic functional

$C(\xi) = \exp\{\int \psi(\xi(t))dt\}$ has independent values at every moment, where ψ is the ψ-function introduced in §3.3.

We introduce a class of subspaces of (L^2) for a general stationary process:

$$L^2(t) = \{\varphi \in (L^2) \ ; \ \varphi \text{ is } \textcircled{B_t}\text{-measurable}\}$$

where $\textcircled{B_t} = \textcircled{B}\{\langle x, \xi \rangle \ ; \ \mathrm{supp}(\xi) \subset (-\infty, t]\}$. Also we introduce the corresponding subspaces of $\mathcal{F}$:

$$\tau(L^2(t)) = \mathcal{F}(t).$$

Obviously we have

$$(17) \qquad \begin{cases} L^2(s) \subset L^2(t) \ , \\ \mathcal{F}(s) \subset \mathcal{F}(t) \ , \quad \text{if} \quad s \leqq t, \end{cases}$$

and

$$(18) \qquad \begin{cases} \bigcup_t L^2(t) = (L^2) \\ \bigcup \mathcal{F}(t) = \mathcal{F} . \end{cases}$$

Definition. A stationary process P is called

a) deterministic if

$$L^2(t) = L^2(s) \text{ for some (hence any) } t,s,$$

b) purely nondeterministic if

(19) $$\bigcap_t L^2(t) = \{1\} \text{ or equivalently } \bigcap_t \mathcal{F}(t) = \{C(\cdot)\}.$$

The following theorem is a generalization of the zero-one law (Theorem 1.2.), therefore we state without proof.

Theorem 4.6. A stationary process with independent values at every moment is purely nondeterministic.

Remark 1. The condition (19) is equivalent to

(19') $$\bigcap \textcircled{B}_t = \{\emptyset,\Omega\}, \text{ mod } 0.$$

We further note that

$$T_t \textcircled{B}_s = \textcircled{B}_{s+t} \text{ for every } t,s.$$

Remark 2. A characteristic functional C with the property

(16) is sometimes called a <u>local</u> functional. Detailed discussions concerning the solution of the functional equation (16) can be found in [16].

Bibliography

[12] R. A. Minlos, Generalized random processes and their extension to a measure. Trudy Moscow Mat. Obšč. 8 (1959), 497-518.

[13] I. M. Gelfand and N. Ya. Vilenkin, Generalized Functions vol. IV. 1961 (English translation: Academic Press).

[14] T. Hida and N. Ikeda, Analysis on Hilbert space with reproducing kernel arising from multiple Wiener integral. Proc. of the 5th Berkeley Symp. on Math. Stat. and Prob. vol. II part I (1967), 117-143.

[15] N. Aronszajn, Theory of reproducing kernels, Trans. Amer. Math. Soc. 68 (1950), 337-404.

[16] M. M. Rao, Local functionals and generalized random fields, Bull. Amer. Math. Soc. 74 (1968), 288-293.

§ 5 Gaussian processes.

We shall first give a definition of Gaussian process. The class o
Gaussian processes is one of the important classes of (generalized) stochastic processes. Then we shall deal with linear operations acting on them. The discussions there will show that the Gaussian white noise plays a dominant role in the study of Gaussian processes.

Throughout this section the nuclear space E will be assumed to b
either the space $\mathcal{S}$ or the space $\mathcal{D}$ (for definitions we refer to Examples i) and ii) in § 4.2) in order to simplify our discussions.

5.1. Definitions of Gaussian process

Let (E^*, μ) be a generalized stochastic process (see § 4.2 Defini
with characteristic functional $C(\xi)$, $\xi \in E$. Assume that

(1) $$\int_{E^*} |< x, \xi >|^2 \, d\mu(x) < \infty \text{ for every } \xi \in E,$$

$$(\text{necessarily } \int_{E^*} | < x, \xi >| \, d\mu(x) < \infty, \ \xi \in E.).$$

Then it is easy to see that

(2) $$m(\xi) \equiv \int_{E^*} < x, \xi > d\mu(x)$$

is a continuous linear functional on E, i.e. $m \in E^*$. Therefore $m(\xi)$ ma
written as $< m, \xi >$. We call m the <u>mean</u> of the stochastic process $(E^*,$

Now consider the covariance $K(\xi, \eta)$ of r.v.s $< x, \xi >$ and $< x, \eta >$

(3) $$K(\xi,\eta) = \int_{E^*} (< x,\xi > - m(\xi)) (< x,\eta > - m(\eta)) \, d\mu(x)$$

$$= B(\xi,\eta) - m(\xi)m(\eta),$$

where $$B(\xi,\eta) = \int_{E^*} < x,\xi > < x,\eta > d\mu(x).$$

The existence of the functionals $K(\xi,\eta)$ and $B(\xi,\eta)$ is guaranteed by the assumption

(1). Obviously $K(\xi,\eta)$ is bilinear and continuous on $E \times E$.

Further $K(\xi,\eta)$ enjoys the property that

$$\sum_{j,k=1}^{n} a_j \cdot \bar{a}_k \, K(\xi_j,\xi_k) \geq 0$$

for any choice of $\xi_1,\ldots,\xi_n \in E$ and of complex numbers $a_1,\ldots,a_n$, that is, K is positive definite. In particular

$$K(\xi,\xi) \geq 0$$

The functional K given by (3) is called the <u>covariance functional</u> of the stochastic process (E^*,μ).

<u>Definition</u> If the characteristic functional $C(\xi)$ of a stochastic process (E^*, μ) is expressed in the form

(4) $$C(\xi) = \exp[\, im(\xi) - \frac{1}{2} K(\xi,\xi)]$$

with a continuous linear functional $m(\xi)$ and a continuous, bilinear, positive definite functional $K(\xi,\eta)$, then the process (E^*,μ) is called <u>Gaussian process.</u>

To give an illustration to the definition of a Gaussian process we observe the following: Let (E^*,μ) be a Gaussian process. Then the r.v. $< x,\xi >$, ξ fixed, has the characteristic function

$$\int_{E^*} e^{iz < x,\xi >} d\mu(x) = \exp(im(\xi)z - \frac{1}{2} K(\xi,\xi)\ z^2),\ z \text{ real.}$$

Therefore $< x,\xi >$ is a Gaussian random variable on (E^*,μ) with mean $m(\xi)$ and variance $K(\xi,\xi)$. Further we see that $< x,\xi >\ ;\ \xi \in E$ is a Gaussian system. In fact, for any finite number of $< x,\xi >$ s, say $< x,\xi_1 >, \ldots, < x,\xi_n >$, we have

$$\int_{E^*} e^{i \sum_{j=1}^{n} z_j < x,\xi_j >} d\mu(x) = \int_{E^*} e^{i < x,\ \Sigma z_j \xi_j >} d\mu(x)$$

$$= \exp[\ i \sum_{j=1}^{n} m(\xi_j) z_j - \frac{1}{2} \sum_{j,k} K(\xi_j,\xi_k) z_j z_k\ ],\ z_1, z_2, \ldots$$

real. This proves that $(< x,\xi_1 >, \ldots, < x,\xi_n >)$ has n-dimensional Gaussian distribution (may be degenerated).

Conversely, if the collection of r.v.s. $< x,\xi >\ ;\ \xi \in E$ forms a Gaussian system, then the characteristic functional can be expressed in the form (4) ; namely we are given a Gaussian process.

5.2. Examples of a Gaussian process.

We shall list some examples of a Gaussian process along with some remarks on them.

Example 1. The stationary process with the characteristic functional

$$C(\xi) = \exp -\frac{1}{2} \int_{-\infty}^{\infty} \xi(t)^2\, dt,\ \xi \in E, \tag{5}$$

is a Gaussian process. The mean and the convariance functional are zero and $\int_{-\infty}^{\infty} \xi(t)\, \eta(t)\, dt$, respectively. This process is called the Gaussian white noise, the detailed discussions of which will be given in Part III. Indeed, the Gaussian white noise is a typical example of a stationary Gaussian process with independent values at every moment.

Example 2. The Brownian motion

Consider the functional

(6) $$C(\xi) = \exp\{-\frac{1}{2}\int_0^\infty |\hat{\xi}(t)|^2 dt\}, \quad \xi \in E,$$

where

$$\hat{\xi}(t) = \int_t^\infty \xi(t)dt, \quad t \geqq 0.$$

Then we can easily see that $C(\xi)$ has an expression of the form (4) with $m = 0$ and $K(\xi,\eta) = \int_0^\infty \hat{\xi}(t)\hat{\eta}(t)dt$. Thus we are given a Gaussian process.

Now observing the following relation

$$\int_0^\infty\int_0^\infty (\min(t, s))\ \xi(t)\ \eta(s)\ dt\ ds = \int_0^\infty \hat{\xi}(t)\ \hat{\eta}(t)dt,$$

we can see that the system $\{ < x,\xi > ;\ \xi \in E\}$ of r.v.s on (E^*, μ) has the same probability distribution as $\{ \int_0^\infty \xi(t)\ B(t,\omega)dt\ ;\ \xi \in E\}$, $\omega \in \Omega(P)$, where $B(t,\omega)$, $t \geq 0$, is the Brownian motion defined in § 2. (Note that if the support of $\xi \subset (-\infty, 0)$, then $< x,\xi > = 0$ a.s. (μ).) We may, therefore, call the Gaussian process (E^*,μ) a Brownian motion.

Example 3. Inspired by the above example we are led to consider a Gaussian process

(7) $$X(t,\omega) = \int_{-\infty}^t F(t-u)\ dB(u) \quad \text{(Wiener integral)}$$

defined on the probability space $(\Omega, \textcircled{B}, P)$. We understand that $F(u) = 0$ for $u < 0$. For simplicity we assume that $X(t,\omega)$ has continuous sample paths. The collection $\{ \int_{-\infty}^\infty \xi(t)\ X(t,\omega)\ dt,\ \xi \in E\}$ forms a Gaussian system. We can prove that

$$\int_{\Omega} \exp\{i \int_{-\infty}^{\infty} \xi(t)\, X(t,\omega)dt\}\, dP(\omega)$$

$$(8) \qquad = \exp\{-\frac{1}{2} \int_{-\infty}^{\infty} (\check{F} * \xi)(t)^2 dt\}, \quad \xi \in E,$$

where $\check{F}(u) = F(-u)$.

Such a consideration shows that starting with the characteristi functional of the form (8) we can discuss, in our set-up, a stationar Gaussian process $X(t)$ given by (7). The Gaussian process (E^*, μ) corresponding to the characteristic functional (8) is a stationary process in our sense.

<u>Example 4.</u> The following characteristic functional

$$(9) \qquad C(\xi) = \exp\{-\frac{1}{2} \int_{-\infty}^{\infty} [(D\xi)(t)]^2 dt\}, \quad \xi \in E, \quad D = -\frac{d}{dt}$$

defines a Gaussian process (E^*, μ). As is easily seen, it is a static process with independent values at every moment.

<u>Example 5.</u> White noise with n-dimensional parameter.

A simple generalization of the Gaussian process leads us to an important class of Gaussian processes with multi-dimensional paramete Now the nuclear space E must be a subspace of $L^2(R^n)$. Functionals $m(\xi)$ and $K(\xi, \eta)$ can be defined in a similar manner to § 5.1.

The following example is of special interest. The functional

$$(10) \qquad C(\xi) = \exp\{-\frac{1}{2} \int_{-\infty}^{\infty} \cdots \int \xi(t_1, \ldots t_n)^2\, dt_1 \ldots dt_n\}$$

defines a Gaussian process which may be called the (Gaussian) white noise with n-dimensional parameter.

5.3. Linear operators acting on Gaussian processes.

Let L be a continuous linear operator on E :

$$L : E \ni \xi \longrightarrow L\xi \in E .$$

Given a Gaussian process (E^*, μ) with the characteristic functional $C(\xi)$, we are given a new functional $C_L(\xi) = C(L\xi)$ by the operator L. Obviously C_L is i) continuous on E, ii) positive definite, and iii) $C_L(0) = 1$ $(L\,0 = 0)$. Hence $C_L(\xi)$ is a characteristic functional. By assumption $C(\xi)$ is expressed in the form (4), and therefore we have

$$C_L(\xi) = \exp\{\, im(L\xi) - \frac{1}{2} K(L\xi, L\xi)\}. \tag{11}$$

We are now given a new Gaussian process (E^*, μ_L) in such a way that

$$C_L(\xi) = \int_{E^*} e^{i\langle x, \xi\rangle} d\mu_L\,(n) = \int_{E^*} e^{i\langle x, L\xi\rangle} d\mu(x)$$
$$= \int_{E^*} e^{i\langle L^* x, \xi\rangle} d\mu(x).$$

The last expression shows that the Gaussian process (E^*, μ_L) is obtained by applying L^* the adjoint of L to the sample function x of the Gaussian process (E^*, μ).

A simple example of the operator L is the differential operator. Set

$$L = D \; (= -\frac{d}{dt}).$$

If we apply the operator $D^* = \frac{d}{dt}$ to the sample function of the Gaussian white noise, then we obtain the Gaussian process given by (9) in Example 4.

Apply D^* to the Brownian motion (Example 2), then we obtain

$$C(D\xi) = \exp\{-\frac{1}{2}\int_0^\infty \xi(t)^2 dt\},$$

which defines the Gaussian white noise restricted to the time domain $[0,\infty)$. In view of this we may roughly say that the Gaussian white noise is obtained by taking the derivative of the Brownian motion.

We shall be able to generalize the above relation between the Brownian motion and the Gaussian white noise by taking a general differential operator L. Let us present an exposition of the simples futures of the general theory to show the idea. Suppose that a Gaussian process is given by the characteristic functional (8). We are interested in finding an operator L such that

(12) $$L^*F = \delta \quad \text{(the delta function)}.$$

and that

(13) $\mathrm{supp}(\xi) \subset (-\infty, a]$ implies that $\mathrm{supp}(L\xi) \subset (-\infty, a]$ for every a, which means that the operator L^* is a so-called causal operator. Suc an operator L^* carries the Gaussian process into the Gaussian white noise in a manner that the value of the white noise at time t is fc by the values of the original process up to time t.

If, in particular, we find a higher order differential operato of the form

(14) $$L = \sum_{k=0}^{n} a_k D^{n-k}, \quad D^0 = 1,$$

which satisfies (12) ((13) is obvious), then the original stationary Gaussian process is an n-ple Markov Gaussian process in the restricted sense.

The converse problem is also interesting for us. Starting with the Gaussian white noise, we shall be able to form a reasonably wide class of purely nondeterministic Gaussian processes by using integral operators satisfying the condition (13).

Such problems stated above and related topics can be discussed for nonstationary Gaussian process. For detailed discussion we refer to P. Lévy's book [5, Complément Chapitre I]. We note that these discussions can be applied to the prediction theory.

Exercise Consider the relation between the differential operator (14) and the kernal function $\check{F}$ appeared in the expression (8). Also, think of the concept of the cannonical representation introduced by P. Lévy (see [5, Complément]) for Gaussian processes.

Part III

Most of this part will be devoted to the detailed discussions of white noise and some of its applications. We refer to Section 4 for background and to Section 2 for illustrations.

§ 6. Hilbert space (L^2) arising from white noise.

6.1. Preliminaries

Let E be a real nuclear space such that

$$E \subset \mathcal{H} = L^2(R^1, \text{Leb}) \subset E^*$$

and such that E is stable under the shifts S_t, $-\infty < t < \infty$. The Gaussian white noise (abbr. W. N.) is a stochastic process with characteristic functional

$$C(\xi) = \exp\{-\frac{1}{2}\int \xi(t)^2 dt\}, \ \xi \in E.$$

Since $C(\xi)$ is S_t-invariant, W.N. is a stationary process. As was seen in the example of § 5.2., W.N. has independent values at every moment. Furthermore we know that W.N. is a Gaussian process and in particular the system $\{\langle x, \xi\rangle ; \xi \in E\}$ is a Gaussian system.

We now proceed to the analysis of $(L^2) = L^2(E^*, \mu)$.

Every assertion in §.4 holds for W.N., and in addition more detailed properties and exact formulas can be established as will be indicated below.

Basic Functions. Exponential functions: Instead of A (see § 4.4) we can take the wider class of exponential functions A' = the algebra spanned by $\{e^{a<x,\xi>};\ \xi \in E$ and a complex$\}$. Obviously $A' \subset (L^2)$ and A' is, of course, dense in (L^2).

Polynomials: We consider the same M as in § 4.4. The vector space M is also dense in (L^2).

The transformation τ. τ is a unitary transformation of (L^2) onto the reproducing kernel Hilbert space $\mathcal{F}$ given by

$$(1) \qquad (\tau\varphi)(\xi) = \int e^{i<x,\xi>}\,\varphi(x)d\mu(x),\ \varphi \in (L^2).$$

We list several formulas below:

$$(L^2) \ni 1 \xrightarrow{\ \tau\ } C(\xi) \qquad \in \mathcal{F}$$

$$e^{i<x,\eta>} \longrightarrow C(\xi + \eta)$$

$$\exp[-i<x,\eta>\sqrt{2}\,t] \longrightarrow C(\xi - \sqrt{2}\,t\eta) = C(\xi)\cdot\exp[\sqrt{2}\,t(\xi,\eta)-t^2$$

$$\text{with } \|\eta\| = 1 \qquad\qquad = C(\xi)\sum_{k=0}^{\infty}\frac{t^k}{k!}\,H_k\left(\frac{(\xi,\eta)}{\sqrt{2}}\right)$$

where $H_k(x)$ is the Hermite polynomial of degree k and $(\cdot,\cdot)$ is the inner product in $L^2(R^1)$. In particular

$$\underset{\text{with } \|\eta\| = 1}{\langle x,\eta\rangle^k} \longrightarrow \left(\frac{i}{\sqrt{2}}\right)^k C(\xi) H_k\left(\frac{(\xi,\eta)}{\sqrt{2}}\right).$$

Further, if $\eta_1, \eta_2, \ldots, \eta_n$ are mutually orthogonal and $\|\eta_j\| = 1$, then

$$\prod_{j=1}^{n} e^{-i\langle x,\eta_j\rangle\sqrt{2}\, t_j} \longrightarrow C(\xi) \prod_{j=1}^{n} e^{\sqrt{2}\, t_j(\xi,\eta_j) - t_j^2}$$

$$(2) \quad \prod_{j=1}^{n} \langle x,\eta_j\rangle^{k_j} \longrightarrow \left(\frac{i}{\sqrt{2}}\right)^{\Sigma k_j} C(\xi) \prod_{j=1}^{n} H_{k_j}\left(\frac{(\xi,\eta_j)}{\sqrt{2}}\right).$$

(For formulas concerning Hermite polynomials we refer to the Appendix.)

The flow $\{T_t\}$. Let T_t be as in § 4.3. The collection $\{T_t, -\infty < t < \infty\}$ forms a flow, i.e. a one-parameter group of measure preserving transformations on $(E^*, Ⓑ, \mu)$. It is called a flow of Brownian motion. By Theorem 4.6, $\{T_t\}$ is a so-called Kolmogorov flow, that is, there exists a sub-σ-field $Ⓑ_o$ of Ⓑ such that, if $Ⓑ_t = T_t Ⓑ_o$

i) $\textcircled{B}_s \subset \textcircled{B}_t$ for $s < t$,

ii) $\bigcap_t \textcircled{B}_t = \{\emptyset, E^*\}$, mod 0,

iii) $\bigcup_t \textcircled{B}_t$ (the σ-field generated by the $\textcircled{B}_t, -\infty < t < \infty$) $= \textcircled{B}$.

The spectral type will be discussed in §7.

6.2. Gauss transform σ.

We introduce another transformation σ on (L^2) which is called the Gauss transform.

Definition. For $\varphi \in (L^2)$ we define $\sigma\varphi$ by

$$(\sigma\varphi)(y) = \int_{E^*} \varphi(ix + y)d\mu(x) \quad , \quad y \in E^*. \tag{3}$$

Notice that $\sigma\varphi$ is a function on E^* while $\tau\varphi$ is a function on E.

Proposition. The Gauss transform σ can be defined on M and A'.

Proof. Suppose $\varphi(x)$ is a monomial, say $\varphi(x) = \prod_{k=1}^{n} \langle x,\xi_k \rangle^{n_k}$. The complexification is well defined: $\varphi(ix+y) = \prod_{k=1}^{n} (i\langle x, \xi_k\rangle + \langle y,\xi_k\rangle)^{n_k}$, $y \in E^*$. This is integrable with respect to $d\mu(x)$ and so we obtain a function $(\sigma\varphi)(y)$ which again belongs to $(L^2)_y$.

(To express the variable explicitly we use the notations such as $(L^2)_x$, M_x, A_x, etc.).

A similar argument works for A'_x .

<u>Example</u> 1. We find the exact form of the Gauss transform of an exponential function $\varphi(x) = e^{\alpha<x,\xi>}$ with $\|\xi\| = 1, \alpha$ complex.

$$(4) \qquad (\sigma\varphi)(y) = e^{\alpha<y,\xi>} \cdot \int_{E^*} e^{i\alpha<x,\xi>} d\mu$$

$$= \exp[\alpha<y,\xi> + \frac{1}{2}(i\alpha)^2].$$

Although σ is linear, the above example shows that σ is not isometric.

If α is real in the expression (4), we obtain the genarating function of Hermite polynomials. Thus we have

$$\sigma(<\cdot,\xi>^n)(y) = \frac{d^n}{d\alpha^n}\sigma(e^{\alpha<\cdot,\xi>})(y)\Big|_{\alpha=0} = 2^{-\frac{n}{2}} H_n\left(\frac{<y,\xi>}{\sqrt{2}}\right)$$

More generally we can prove the following

$$(5) \quad \sigma\left(\prod_{j=1}^{n} <\cdot,\xi_j>^{k_j}\right)(y) = 2^{-\frac{1}{2}\Sigma k_j} \prod_{j=1}^{n} H_{k_j}\left(\frac{<y,\xi_j>}{\sqrt{2}}\right), \quad (\xi_j,\xi_k) = \delta_{j,k}$$

A function expressed in the form of the right hand side of (5) is called a Fourier-Hermite polynomial (of degree Σk_j).

Theorem 6.1. Let M_n be the collection of all the homogeneous polynomials of degree n. Then $M = \sum_{n=0}^{\infty} M_n$ and the $\sigma(M_n)$, $n = 0, 1, \ldots$, are vector spaces $(\subset (L^2))$ satisfying

i) $\qquad \sum_{k=0}^{n} \sigma(M_k) = \sum_{k=0}^{n} M_k$ for every n,

ii) $\qquad \sigma(M_n) \perp \sigma(M_m)$ if $n \neq m$.

Proof. i) is obvious.

For the proof of ii) it suffices to show that $\varphi(x)$ and $\psi(x)$ expressed in the form

$$\varphi(x) = \prod_j H_{k_j}\left(\frac{\langle x, \xi_j \rangle}{\sqrt{2}}\right), \quad \Sigma k_j = n,$$

$$\psi(x) = \prod_j H_{\ell_j}\left(\frac{\langle x, \xi_j \rangle}{\sqrt{2}}\right), \quad \Sigma \ell_j = m,$$

by the same orthonormal system $\{\xi_j\}$ (in $L^2(R^1)$) are mutually orthogonal. But the orthogonality property of the Hermite polynomials with respect to the Gaussian measure proves the assertion.

Let $\mathcal{H}_n$ be the closure (in (L^2)) of $\sigma(M_n)$. Then, noting that M is dense in (L^2), Theorem 6.1 implies the following

Corollary. (Wiener's direct sum decomposition)

$$(6) \qquad (L^2) = \sum_{n=0}^{\infty} \oplus \mathcal{H}_n , \quad \mathcal{H}_n = \overline{\sigma(M_n)}$$

Definition. Each element of $\mathcal{H}_n$ is called a multiple Wiener integral of degree n.

6.3. Realization of the multiple Wiener integral.

We start with the proof of the following

Proposition. Let $\{\xi_j\}$ be an arbitrarily fixed c.o.n.s. in $L^2(R^1)$ such that each ξ_j is in E. Then Fourier-Hermite polynomials of the form

$$(7) \qquad \prod_j c_{k_j} H_{k_j}\left(\frac{\langle x, \xi_j\rangle}{\sqrt{2}}\right) \quad \text{with} \quad \Sigma k_j = n$$

form a c.o.n.s. in $\mathcal{H}_n$, where $c_k = (\sqrt{k!2^k})^{-1}$.

Proof. First we note that any ξ in E can be approximated (in $L^2(R^1)$) by a linear combination of the ξ_j. Hence, by using the addition formula for Hermite polynomials (see Appendix), we can prove that any Fourier-Hermite polynomial of degree n is approximated (in (L^2)) by the linear combination of the Fourier-Hermite polynomials based on the ξ_j of the same degree n. Thus we have proved that the functions of the form (7) span the subspace $\mathcal{H}_n$.

The rest of the proof is obvious.

Now we establish a nice realization of $\mathcal{H}_n$. The idea comes from K. Ito [17], S. Kakutani [18], and N. Wiener [19].

[I] First let us note that $\mathcal{H}_1$ is nothing but the Wiener integral introduced in §2.3. Indeed, the correspondence

$$\mathcal{H}_1 \ni < x, \xi > \longleftrightarrow \xi \in L^2(R^1)$$

extends to a one to one isometric mapping of $\mathcal{H}_1$ onto $L^2(R^1)$. Thus $L^2(R^1)$ can be thought of as a realization of H_1 :

$$(8) \qquad \mathcal{H}_1 \cong L^2(R^1).$$

Another exceptional case is the case $n = 0$. Obviously

$$(9) \qquad \mathcal{H}_0 \cong R^1.$$

For general $\mathcal{H}_n$, $n > 1$, we can take $\widehat{L^2}(R^n) = \{F; F \in L^2(R^n)$ and F is symmetric$\}$. The situation is illustrated in the following manner. Let $\{\xi_j\}$ be a fixed c.o.n.s. for $L^2(R^1)$. By Proposition the Fourier-Hermite polynomials of the form (7) span $\mathcal{H}_n$, therefore it suffices to establish a realization for Fourier-Hermite polynomials. Since the polynomial given by (7) is determined by the n-tuple of ξ_j's, it is quite reasonable to consider a mapping

$$(10) \quad I_n': \xi_{p_1}(t_1)\,\xi_{p_2}(t_2)\cdots\xi_{p_n}(t_n) \longrightarrow \prod_j c_{k_j} H_{k_j}\left(\frac{\langle x, \xi_j\rangle}{\sqrt{2}}\right)$$

where $(p_1, p_2, \ldots, p_n)$ is an n-tuple of positive integers and k_j is the number of the p_i such that $p_i = j$. The norm (one is $L^2(R^n)$-norm and the other is (L^2)-norm) is preserved under the mapping I_n'. However I_n' is a <u>many</u> to one mapping. Therefore we introduce an equivalence relation to $L^2(R^n)$ so that the factor space turns out to be isomorphic to $\mathcal{H}_n$. Let π denote a permutation of $(1,2,\ldots,n)$, and define

$$f^\pi(t_1,t_2,\ldots,t_n) = f(t_{\pi(1)},t_{\pi(1)},t_{\pi(2)},\ldots,t_{\pi(n)}).$$

Then, by definition, we have

$$I_n'(f^\pi) = I_n'(f)$$

for a function appeared in (10). Denote by $\tilde{f}$ the symmetrization of f: $\tilde{f} = \frac{1}{n!}\sum_\pi f^\pi$. The equivalence $f_1 \sim f_2$ means that $\tilde{f}_1 = \tilde{f}_2$. Now it is easy to see that

(11) $\qquad I'_n(f_1) = I'_n(f_2)$ if and only if $f_1 \sim f_2$

for particular functions expressed as linear combinations of the $\prod_{i=1}^{n} \xi_{p_i}(t_i)$. The mapping I'_n extends to a one to one mapping from $L^2(R^n)/\sim$ onto $\mathcal{H}_n$. Since the factor space $L^2(R^n)/\sim$ is isomorphic to $\widehat{L^2}(R^n)$, we finally obtain an isometry $I_n : \widehat{L^2}(R^n) \longrightarrow \mathcal{H}_n$. Elementary computations, although they are somewhat complicated, enable us to prove

<u>Theorem</u> 6.2. The mapping $I_n : \widehat{L^2}(R^n) \longrightarrow \mathcal{H}_n$ is one to one, onto and linear. Moreover

(12) $$\|I_n(F_n)\|_{(L^2)} = \sqrt{n!}\, \|F_n\|_{L^2(R^n)}, \qquad F_n \in \widehat{L^2}(R^n).$$

Set $I = \sum_{n=0}^{\infty} \oplus I_n$, I_0 = identity, and I_1 = the mapping given by (8).

<u>Corollary</u>. The mapping I gives a linear isomorphism

$$(L^2) \cong \sum_{n=0}^{\infty} \oplus \widehat{L^2}(R^n)$$

in such a way that for $F = \{F_n\}$, $F_n \in \widehat{L^2}(R^n)$, $I(F) = \sum_{n=0}^{\infty} I_n(F_n)$ satisfies

$$(12') \qquad \|I(F)\|^2_{(L^2)} = \sum_{n=0}^{\infty} n!\, \|F_n\|^2_{L^2(R^n)}.$$

[II] We consider another approach to the realization of the multiple Wiener integral, using the transformation τ. After establishing the formula (cf. the formula (2))

$$\tau(\exp[2t\cdot \frac{<\cdot,\eta>}{\sqrt{2}} - t^2])(\xi) = C(\xi) \sum_{k=0}^{\infty} \frac{(\sqrt{2}it)^k}{k!} (\eta,\xi)^k .$$

With $\|\eta\| = 1$, we obtain

$$(13) \qquad \tau\left(H_k\left(\frac{<\cdot,\eta>}{\sqrt{2}}\right)\right)(\xi) = (\sqrt{2}\, i)^k\, C(\xi)(\int \eta(t)\xi(t)dt)^k,$$

and hence we have for an orthonormal system $\{\eta_j\}$,

$$(14) \qquad \tau\left(\prod_{\Sigma k_j=n} c_{k_j} H_{k_j}\left(\frac{<\cdot\eta_j>}{\sqrt{2}}\right)\right)(\xi)$$

$$= i^n C(\xi)(\Pi \mathbf{k}_j)^{-\frac{1}{2}} \int_{R^n}\!\!\cdots\!\int \eta_1(t_1)\cdots\eta_1(t_{k_1})\eta_2(t_{k_1+1})\cdots\xi(t_1)\cdots\xi(t_n)(dt)^n.$$

In the integral in (14) η_j appears as many times as k_j. Now we proceed as in Case [I]; we symmetrize the integrand in (14) without destroying the equality. Then

$$(14') \qquad \tau\left(\prod_{\Sigma k_j = n} c_{k_j} H_{k_j}\left(\frac{\langle \cdot\, \xi_j \rangle}{\sqrt{2}}\right)\right)(\xi)$$

$$= i^n C(\xi) \int \ldots \int F_n(t_1,\ldots,t_n)\, \xi(t_1) \ldots \xi(t_n)\, dt_1 \ldots dt_n$$

where $F_n \in \widehat{L^2}(R^n)$ and $\|F_n\|_{L^2(R^n)} = (n!)^{-\frac{1}{2}}$ which can be generalized to a relation

$$(15) \qquad \mathcal{H}_n \ni \varphi(x) \longleftrightarrow F_n \in L^2(R^n)$$

such that

$$\tau(\varphi)(\xi) = i^n C(\xi) \int_{R^n}\!\!\ldots\int F_n(t_1,\ldots,t_n)\, \xi(t_1) \ldots \xi(t_n)\, dt_1 \ldots dt_n$$

and such that

$$\|\varphi\|_{(L^2)} = \sqrt{n!}\, \|F_n\|_{L^2(R^n)}.$$

Thus we have the same realization of $\mathcal{H}_n$.

Bibliography

[17] K. Itô, Multiple Wiener integral. J. Math. Soc. Japan, vol. 3 (1951), 157-169.

[18] S. Kakutani, Spectral analysis of stationary Gaussian processes. Proc. 4th Berkeley Symp. 1961, vol. 2, 239-247.

[19] N. Wiener, Nonlinear problems in random theory. M.I.T., Wiley, 1958.

§7. Flow of the Brownian motion.

Let μ on $(E^*, \textcircled{B})$ be W. N. (white noise) with the characteristic functional

$$C(\xi) = \exp\{-\tfrac{1}{2}\int \xi(t)^2 dt\}, \quad \xi \in E.$$

Since W. N. is stationary, $\{T_t; t \text{ real}\}$ is a flow on (E^*, μ). According to the discussion in §4.3, the U_t defined by

$$U_t\varphi(x) = \varphi(T_t x), \quad \varphi \in (L^2),$$

form a strongly continuous one parameter group of unitary operators acting on (L^2). Therefore we can appeal to Stone's theorem which asserts that $\{U_t\}$ has a spectral decomposition:

$$(1) \qquad U_t = \int e^{it\lambda}\, dE(\lambda),$$

where $\{E(\lambda); \lambda \text{ real}\}$ is a resolution of the identity.

We are interested in the spectral type of $\{U_t\}$ (or $\{T_t\}$) which we study using the multiple Wiener integral and its realization (cf. K. Ito [20], S. Kakutani [21]).

In order to illustrate the "spectral type", we first state Hellinger-Hahn's theorem. For detailed discussions we refer to Halmos [21].

<u>Theorem</u>. (Hellinger-Hahn) Let $\{U_t;\ t \text{ real}\}$ be a strongly continuous one-parameter group of unitary operators acting on a separable Hilbert space H with the representation (1). Then H is the direct sum of two subspaces:

$$H = M \oplus N, \tag{2}$$

where M and N satisfy the following properties.

i) U_t has discrete spectrum on M, namely M is the direct sum of one-dimensional subspaces M_k on which U_t acts in such a way that $U_t g = \exp[it\lambda_k]g$, $g \in M_k$, $-\infty < t < \infty$. ($\{\lambda_k\}$ is the discrete spectrum of $\{U_t\}$).

ii) N is the direct sum of cyclic subspaces N_k, that is, there exist vectors f_k in H such that N_k is spanned by the $U_t f_k$, $-\infty < t < \infty$, and $N = \sum_k \oplus N_k$. If we set $\rho_k(\lambda) = \|E(\lambda)f_k\|^2$, then $\rho_k(\lambda)$ is continuous and

$$d\rho_1 \gg d\rho_2 \gg \cdots .$$

Furthermore M_k is isomorphic to $L^2(R^1, d\rho_k)$ and U_t restricted to M_k is equivalent to V_t on $L^2(R^1, d\rho_k)$ defined by

(3) $\qquad (V_t g)(\lambda) = \exp[it\lambda]g(\lambda) \quad , \quad g \in L^2(R^1, d\rho_k).$

iii) The subspaces M and N are uniquely determined. If another decomposition $M = \Sigma \oplus M'_k$ and $N = \Sigma \oplus N'_k$ is given, then $\{\lambda'_k\}$ corresponding to the M'_k is exactly the same as $\{\lambda_k\}$ including multiplicity and the measures $d\rho'_k$ associated with the N_k by ii) satisfy $d\rho'_k \sim d\rho_k$ (equivalent) for every k.

The λ_k with multiplicity and the type of the $d\rho_k$ are called the <u>spectral type</u> of $\{U_t\}$. The spectral type determines $\{U_t\}$ up to unitary equivalence. Indeed, let $\{U_t\}$ on H and $\{U'_t\}$ on H' be one-parameter unitary groups and let T be a unitary transformation of H onto H' such that $U'_t = TU_tT^{-1}$. Then $\{U_t\}$ and $\{U'_t\}$ have the same spectral type. The converse is also true. Thus up to unitary equivalence, the spectral type classifies one parameter unitary groups.

<u>Example</u>. Let $H = L^2(R^1, m)$, m = Lebesgue measure, and let U_t be given by $U_t f(x) = f(x-t)$, $f \in H$. The Hilbert space H is a cyclic subspace for U_t, e.g. if we take f_1 such that the Fourier transform does not vanish, then $\{U_t f_1; -\infty < t < \infty\}$ spans the whole space H. Hence we see that $M = \{0\}$ and $N = N_1$

with $d\rho_1 \sim m$. In this case we say that $\{U_t\}$ has a simple Lebesque spectrum.

If $M = \{0\}$, and if every $d\rho_k$ is equivalent to the Lebesgue measure or 0, then $\{U_t\}$ is said to have Lebesgue spectrum. In particular, if there exist infinitely many $d\rho_k$'s equivalent to the Lebesgue measure, we say that $\{U_t\}$ has ρ-Lebesgue spectrum.

Our purpose is to prove that $\{U_t\}$ derived from the flow of the Brownian motion has σ-Lebesgue spectrum on $(L^2) \ominus \{1\} = \sum_{n=1}^{\infty} \oplus \mathcal{H}_n$, where $\mathcal{H}_n$ is the multiple Wiener integral. First we note the following

Proposition

$$U_t \mathcal{H}_n = \mathcal{H}_n \quad \text{for every } n. \tag{4}$$

Hence we can determine the spectral type of $\{U_t\}$ on each $\mathcal{H}_n$ separately. We now use the realization of $\mathcal{H}_n$ discussed in §6.3. [II]. By the transformation τ, U_t on (L^2) is transformed to a unitary group $\tilde{U}_t$ on $\mathfrak{F}$ in such a way that

$$\tau(U_t(\varphi(x))(\xi) = \tau(\varphi(T_t x))(\xi)$$
$$= \int e^{i<x,\xi>} \varphi(T_t x)d\mu(x) = \int e^{i<T_{-t}x',\xi>} \varphi(x')d\mu(x'), x' = T_t x$$
$$= \tau(\varphi)(S_{-t}\xi) \equiv \tilde{U}_t(\tau(\varphi))(\xi).$$

Therefore, in the correspondence between $\mathcal{H}_n$ and $L^2(R^n)$ given by the formula (15) of §6.3, if

$$\tau(\varphi)(\xi) = i^n C(\xi) \int \underset{R^n}{\cdots} \int F_n(t_1,\ldots,t_n)\xi(t_1) \ldots \xi(t_n)dt^n, \ \varphi \in \mathcal{H}_n,$$

then we have

$$\tilde{U}_t(\tau(\varphi))(\xi) = i^n C(\xi) \int \underset{R^n}{\cdots} \int F_n(t_1,\ldots,t_n)\xi(t_1+t) \ldots \xi(t_n+t)dt^n.$$

In other words, if $\varphi \in \mathcal{H}_n$ corresponds to $F_n \in L^2(R^n)$, then we associate $F_n(t_1-t,\ldots,t_n-t)$ with $U_t\varphi$.

Our problem is now to find the spectral type of the following one parameter unitary group $\{U_t^{(n)}\}$:

$$U_t^{(n)}F_n(t_1,\ldots,t_n) = F_n(t_1-t,\ldots,t_n-t) \ , \ F_n \in L^2(R^n).$$

Changing the coordinate system of R^n from $(t_1,\ldots,t_n)$ to $(u_1,\ldots,u_n)$ so that

$$u_1 = \frac{1}{n}\sum_{j=1}^{n} t_j,$$

$u_2,\ldots,u_n$ are linear combinations of the t_j and the system is invariant under the shift $t_j \longrightarrow t_j - t$, and that the Jacobian $\dfrac{\partial(u_1,\ldots,u_n)}{\partial(t_1,\ldots,t_n)} = 1$, we see that $F(t_1,\ldots,t_n)$ is expressed as $f(u_1,u_2,\ldots,u_n)$ and that

$$U_t^{(n)}F(t_1,\ldots,t_n) = f(u_1-t,\ u_2,\ldots,u_n).$$

Now we introduce a C.O.N.S. $\{\xi_n\}$ for $L^2(R^1)$ such that the Fourier transform of every ξ_n never vanishes. Then we have the Fourier series expansion of f:

$$f(u_1,\ldots,u_n) = \Sigma\, a_{k_1 \ldots k_n}\, \xi_{k_1}(u_1)\xi_{k_2}(u_2) \ldots \xi_{k_n}(u_n)$$

and we have

$$(5)\quad U_t^{(n)}f(u_1,\ldots,u_n) = \Sigma\, a_{k_1 \ldots k_n}\, \xi_{k_1}(u_1-t)\xi_{k_2}(u_2) \ldots \xi_{k_n}(u_n).$$

By the formula (5), we see that the $U_t^{(n)}f$, $-\infty < t < \infty$, span a subspace $L_{k_2k_3 \ldots k_n}$ of $\widehat{L^2}(R^n)$ which is isomorphic to $L^2(R^1)$, and on which $U_t^{(n)}$ acts like U_t in the above example. Therefore $U_t^{(n)}$ has simple Lebesgue spectrum on $L_{k_2 \ldots k_n}$. It is easy to see that $\widehat{L^2}(R^n)$ is a direct sum of countably infinite number of subspaces of this kind unless $n = 1$. The exceptional case where $n = 1$ is the same as in the above example. Thus we have proved

<u>Theorem</u>. The unitary group $\{U_t\}$ derived from the flow of Brownian motion has

i) simple Lebesgue spectrum on $\mathcal{H}_1$, and

ii) σ-Lebesgue spectrum on $\mathcal{H}_n$, $n \geq 2$.

[Bibliography]

[20]. K. Itô, Spectral type of the shift transformation of differential processes with stationary increments. Trans. AMS vol. 81 (1956), 253-263.

[21]. S. Kakutani, Determination of the spectrum of the flow of Brownian motion, Proc. Nat. Acad. Sci. USA vol. 36 (1950), 319-323.

[22]. P. R. Halmos, Introduction to Hilbert space and the theory of spectral multiplicity. N.Y., 1957.

§8. Infinite dimensional rotation group

After H. Yosizawa we consider the collection $O_\infty(E)$ of all linear transformations $\{g\}$ on E satisfying the following two conditions:

$$\text{(1)}\quad \begin{cases} \text{i)} \ g \text{ is a homeomorphism on } E, \\ \text{ii)} \ \int (g\xi)(t)^2 dt = \int \xi(t)^2 dt, \end{cases}$$

where E is a nuclear space dense in $L^2(R^1)$. Obviously the set $O_\infty(E)$ of all such g's forms a group with respect to the operation $(g_1g_2)\xi = g_1(g_2\xi)$. The group $O_\infty(E)$ is called the infinite dimensional rotation group. Since the characteristic functional of W.N. is invariant under $O_\infty(E)$, the measure μ of W.N. is invariant under g^* the adjoint of g, $g \in O_\infty(E)$.

8.1. A subgroup arising from time change.

We first investigate a subgroup $G(\subset O_\infty(E))$, each member g of which acts in such a way that

$$(g\xi)(\varphi) = \xi(\psi(\varphi))\sqrt{|\psi'(\varphi)|}, \tag{2}$$

where $\psi(\varphi)$ is a strictly monotone function. Of course, we require some regularity conditions for ψ so that the condition (1),i) is satisfied. Such conditions depend on the choice of the nuclear space E. It is quite easy to see that G forms a subgroup of $O_\infty(E)$ as soon as E is specified.

We are interested in one parameter subgroups $\{g_t\}$ of G determined by a family $\{\psi_t(\varphi)\}$:

$$g_t : \xi(\varphi) \longrightarrow \xi(\psi_t(\varphi))\sqrt{\left|\frac{d}{d\varphi}\psi_t(\varphi)\right|},$$

where the ψ_t satisfy

(3) $$\psi_t \circ \psi_s(\varphi) = \psi_{t+s}(\varphi) \quad \text{for every } t \text{ and } s.$$

By Aczél [24, chap. 6], $\psi_t(\varphi)$ can be expressed in the form

(4) $$\psi_t(\varphi) = f[f^{-1}(\varphi) + t]$$

with arbitrary f continuous and strictly monotonic.

We now assume that f in the expression (4) satisfies certain an. properties so that $g_t \in O_\infty$ for every t.

<u>Example 1.</u> Let $E = \mathcal{L}$ and set $\psi_t(\varphi) = \varphi - 5$. Then g_t is the shift S_t

<u>Example 2.</u> Let ψ_t be given by $\psi_t(\varphi) = \frac{\varphi}{1-t\varphi}$. Obviously the ψ_t satisfy the condition (3), although each ψ_t has a discontinuity. It is a great restriction on the nuclear space E to assume that it is stable under g_t. The space D_o given in the following is an example which is stable under g_t (for details we refer to Gelfand-Graev-Vilenkin [23, Chapt. VII.]).

$$D_o = \{\xi(\varphi) \; ; \; \xi \in C^\infty , \; \hat{\xi} \in C^\infty\}, \quad \text{where}$$
$$\hat{\xi}(\varphi) = \xi(-\tfrac{1}{\varphi}) \; \frac{1}{|\varphi|}$$

Associated with $\{g_t\}$ is the infinitesimal generator A:

$$A = \frac{d}{dt} \, g_t \Big|_{t=o} .$$

If $\{g_t\}$ is determined by ψ_t of the form (4), simple computations lead us to

$$(A\xi)(\varphi) = f'[f^{-1}(\varphi)] \frac{d}{d\varphi} \xi(\varphi) + \frac{1}{2} \{f'[f^{-1}(\varphi)]\}' . \xi(\varphi).$$

Setting $\quad a(\varphi) = f'[f^{-1}(\varphi)] = \{[f^{-1}(\varphi)]'\}^{-1}$ we have

(5) $$A = a(\varphi)\frac{d}{d\varphi} + \frac{1}{2} a'(\varphi).$$

Thus the infinitesimal generator of any one parameter subgroup $\{g_t\}$ of G is determined by a single function $a(\varphi)$, which is determined by f in (4). The generator given by (5) is denoted by A^a when many generators are discussed. The following equation is straightforward.

(6) $$[A^a, A^b] = A^c, \qquad c = ab' - a'b,$$

where $[\ ,\ ]$ denotes the commutator, i.e. $[A,B] = AB - BA$.

Now our problems in this section are stated as follows.

[1] Find possible one parameter subgroups of G with generator given by (5).

[2] Observe the roles of each one parameter subgroup in the theory of probability.

8.2. Shift.

Set $f(\varphi) = -\varphi$, then $a(\varphi) = -1$ and we obtain $A = -\frac{d}{d\varphi}$ which we denote by $\mathcal{s}$. The generator $\mathcal{s}$ is associated with the shift $\{S_t\}$ by which stationarity is defined. The adjoint $\{T_t\}$ of the shift is the flow of the Brownian motion discussed in §7. Therefore it is quite reasonable to start with $\{S_t\}$.

8.3. Tension. (Dilation)

We shall find a one parameter group $\{g_t\}$ nicely related to the shift.

First of all we consider the case where $\{g_t\}$ is commutative with $\{S_t\}$, namely,

(7) $$[A, \mathcal{s}] = 0$$

where A is the infinitesimal generator of $\{g_t\}$. If $A = A^a = a\frac{d}{d\varphi} + \frac{1}{2}a'$, then (7) implies that $a' \frac{d}{d\varphi} + \frac{1}{2} a'' \equiv 0$. Hence a = constant, i.e. A is a constant multiple of $\mathcal{L}$, therefore $\{g_t\}$ is essentially the sam as $\{S_t\}$.

Another possible relation between A and $\mathcal{L}$ is

(8) $$[A, \mathcal{L}] = \lambda \mathcal{L}, \quad \lambda \text{ constant}$$

This relation shows that $f(\varphi) = -(\exp \lambda \varphi)/_\lambda$ and hence

$$\psi_t(\varphi) = \varphi \exp(\lambda t).$$

Without loss of generality we can assume $\lambda = 1$. Then the generator of the form $\varphi \frac{d}{d\varphi} + \frac{1}{2}$ which will be denoted by h. Setting $H_t = \exp\{th\}$, we have

(9) $$H_t S_s = S_{se^t} H_t.$$

In terms of the theory of dynamical system, the relation (9) shows that the flow $\{T_t\}$ with $T_t = S_t^*$ is transversal flow to the flow $\{H_t^*\}$. We also refer to the tension group acting on Lévy processes discussed in §3.4. and §3.5.

Remark 1. The flow $\{H_t^*\}$ is equivalent to the flow of Ornstein-Uhlen Brownian motion in the following sense. Set $U(t,x) = \langle H_{-t}^* x, I_k \rangle$, where I_k is the indicator function of the unit interval K. note tha $U(t,x)$ makes sense although $I_k \notin E$ (cf §6.3 [1]). It is easy to see that $U(t,x)$, $-\infty < t < \infty$, is a Gaussian process with covariance function $\exp\{-|h|/2\}$, that is $U(t)$ is the Ornstein - Uhlenbeck Brownian motion.

Remark 2. The $U(t)$ defined above has the following canonical representation in terms of Brownian motion $B(t)$ (see §5).

$$U(t) = \int_{-\infty}^{t} \exp\{-(t-u)/2\} dB(u). \tag{10}$$

On the other hand, by a formal differentiation, we have

$$\frac{d}{dt} U(t) = \langle x, \frac{d}{dt} H_{-t} I_k \rangle = \langle x, H_{-t}(\varphi \delta_1(\varphi)) - \frac{1}{2} H_{-t} I_k \rangle ,$$

which implies the so-called Langevin equation

$$dU(t) = d\bar{B}(t) - \frac{1}{2} U(t) dt$$

where $\bar{B}$ and B in (10) are related by $e^{t/2}\bar{B}'(e^t) = B'(t)$.

8.4. $PLG(2,\mathbb{R})$.

We are interested in finding another one parameter subgroup of G which has its own probabilistic meaning and is nicely related to the shift and the tension. First we are concerned with infinitesimal generators. Suppose $A = A^a$ is the generator of the one parameter group which we are looking for. If we require that the system $\{\ ,h,A\}$ is closed under the operation $[\ ,\ ]$, then the relations

$$[\boldsymbol{s}, A] = -a'(\varphi)\frac{d}{d\varphi} - \frac{1}{2} a''(\varphi)$$
$$[h,A] = (\varphi a'(\varphi) - a(\varphi)) \frac{d}{d\varphi} + \frac{1}{2} \varphi \,.\, a''(\varphi)$$

tell us that $a'(\varphi)$ and $\varphi a'(\varphi) - a(\varphi)$ must be linear combinations of 1, φ, and $a(\varphi)$. This requirement implies that $a(\varphi) = c\varphi^2 + d\varphi + e$, c, d, e constants. Therefore as a basis we can choose $\{\boldsymbol{s}, h, k\}$, where $k = \varphi^2 \frac{d}{d\varphi} + \varphi$. The operator k is the infinitesimal generator of the one parameter group $\{K_t\}$ given by

$$K_t\xi(\varphi) = \xi\left(\frac{\varphi}{1-t\varphi}\right)\frac{1}{1-t\varphi}, \qquad -\infty < t < \infty . \tag{11}$$

Observe that the relations

$$\begin{aligned} [\mathscr{s},h] &= -\mathscr{s}, \\ [\mathscr{s},k] &= -2h, \\ [h,k] &= k \end{aligned} \tag{12}$$

are the same as the relations which are satisfied by the Lie algebra of the group $SL(2,\mathbb{R})$ up to multiplicative constants.

Another direct motivation for introducing $\{K_t\}$ is the followir As we discussed in §3.5, the shift and the tension are isomorphic to t groups $N = \{\begin{pmatrix}1 & b\\ 0 & 1\end{pmatrix}\}$ and $A = \{\begin{pmatrix}a & 1\\ 0 & a^{-1}\end{pmatrix}\}$ respectively. They arise in the Bruhat decomposition of $SL(2,\mathbb{R})$:

$$SL(2,\mathbb{R}) = N\ A \cup N\ A\ s\ N,$$

where $s = \begin{pmatrix}0 & 1\\ -1 & 0\end{pmatrix}$. Thus it is quite reasonable to think of a subgrou sNs^{-1} which corresponds to $\{K_t\}$.

Now we come to a very important question which asks if there ex a nuclear space E which is stable under the S_t, H_t and K_t. The answer is given by Example 2, that is, we can take the space D_0. Take D_0 and introduce a trivial transformation.

$$V : \overset{V}{\xi(\varphi)} = \xi(-\varphi). \tag{13}$$

Then we obtain a subgroup G_0 of $O_\infty(D_0)$ which is isomorphic to $PLG(2,\mathbb{R})$:

G_0 = the group generated by the S_t, H_t, K_t and V.

Finally a probabilistic interpretation is given for the group $\{K_t\}$. In fact, Lévy's projective invariance for the Gaussian process $\xi(t)$, $t_1 \le t \le t_2$, introduced in §2.1. corresponds in the present set up to the fact that G_o, and in particular $\{K_t\}$, is a subgroup of O_∞. The proof is left to the reader, although the following example gives some idea of the proof.

Example 3. Let $t_1 = 0$ and $t_2 = 1$. Then $\xi(t)$, $0 \leq t \leq 1$, has the following canonical representation (see §5):

$$\xi(t) = \sqrt{\frac{1-t}{t}} \int_0^t \frac{1}{1-u} dB(u).$$

Three concluding remarks are in order.

Remark 3. The group G_o can not extend so that the Lie algebra is generated by A^a With polynomial a. In this sense G_o satisfies a maximal property.

Remark 4. If we introduce a c. o. n. s. $\{\xi_n\}$ for $L^2(R^1)$ with $\xi_n \in E$, we can discuss finite dimensional rotations. They form an important subgroup of O_∞. Such a subgroup will be useful in the determination of the infinite dimensional Laplacian operator.

Remark 5. The groups such as $SL(n,\mathbb{R})$, $SL(2,C)$, etc. arise in a similar approach.

[Bibliography]

[23] I.M. Gelfand, M.I. Graev and N. Ya. Vilenkin, Generalized Functions, vol. 5. (English translation, Academic Press. 1966)

[24] J. Aczél, Functional Equations and their applications. Academic Press 1966.

§9. Fourier Analysis on (L^2), motion group and Laplacian.

We have already discussed the transformation τ. This has a formal resemblance to the ordinary Fourier transform, but it differs crucially in that it maps the (L^2) not onto itself but to the reproducing kernel Hilbert space $\mathcal{F}(E,C)$. We first discuss anothe of transformation which does play the role of Fourier transform on (L^2). We then proceed to the infinite dimensional motion group and observe the intimate relation with this Fourier transform. Finally we introduce the infinite dimensional Laplacian, which is discussed connection with O_∞ and the direct sum decomposition of (L^2).

9.1. Fourier-Wiener transform.

Let $\varphi(x)$ be in $(L^2) = L^2(E^*,\mu)$. We define

i) $\sqrt{2}$-complexification: $\varphi(x) \longrightarrow \varphi(\sqrt{2}\, x+iy)$, $y \in E$

ii) Fourier-Wiener transform F:

$$\varphi(x) \longrightarrow (F\varphi)(y) = \int_{x \in E^*} \varphi(\sqrt{2}\, x+iy)\, d\mu(x) . \tag{1}$$

Remark 1. The definition of the Fourier-Wiener transform is d to Cameron and Martin [25]. See also N. Wiener [26].

Proposition 1. The Fourier-Wiener transform F is defined on M and A, and F maps them onto themselves. (See §4.4 for notati

Proof. Let $\varphi(x)$ be a Fourier-Hermite polynomial, say

$$\varphi(x) = \prod_k H_{n_k}\left(\frac{\langle x, \xi_k \rangle}{\sqrt{2}}\right), \text{ where } \{\xi_k\} \text{ is a c.o.n.s. for } L^2(R^1)$$

The $\sqrt{2}$-complexification makes sense:

$$\varphi(\sqrt{2}\ x+iy) = \prod_k H_{n_k}\left(< x,\xi_k> + i\ \frac{< y,\xi_k>}{\sqrt{2}}\right) .$$

Since the $< x,\xi_k >$ are mutually independent, the formula

$$H_n(\alpha x + \beta y) = \sum_{j=0}^{n} \binom{n}{j} \alpha^{n-j}\beta^j H_{n-j}(x)H_j(y) , \quad \alpha^2 + \beta^2 = 1,$$

yields

$$\int_{x\in E^*} \varphi_*(\sqrt{2}\ x+iy)d\mu(x) = \prod_k \int_{x\in E^*} \left[\sum_{j=0}^{n_k} \binom{n_k}{j} (\sqrt{2})^{n_k-j}(i)^j \right.$$

$$\left. H_{n_k-j}\left(\frac{< x,\xi_k >}{\sqrt{2}}\right) H_j\left(\frac{< y,\xi_k >}{\sqrt{2}}\right)\right] d\mu(x) ,$$

or

(2) $$(F\varphi)\ (y) = (i)^{\Sigma n_k}.\varphi(y) .$$

This expression shows that the transformation F is defined on M and that

(3) $$F(M_x) = M_y .$$

For an exponential function $\varphi(x)$, say $\varphi(x) = \exp[a < x,\xi >]$ with complex a and $\|\xi\| = 1$, we can easily see that the transformation F is defined and that

(4) $$(F\varphi)\ (y) = \exp[ia < y,\xi > + a^2] .$$

(5) $$F(A_x) = A_y .$$

Theorem 9.1. The transformation F extends uniquely to a unitary operator on (L^2). On the space H_n the multiple Wiener integral of degree n, F acts in such a way that

(6) $$F\varphi(x) = i^n\varphi(x) , \quad \varphi \in H_n .$$

The theorem is easily proved using the relation (2) and the fact that M is dense in (L^2). As an immediate consequence of the theorem we have

$$(7) \qquad F^2 = -I \quad (I \text{ is the identity}).$$

<u>Remark 2</u>. We can give a plausible interpretation to the $\sqrt{2}$-complexification. Let F_a, a complex, be a transformation given by

$$(F_a\varphi)(x) = \int_{E^*} \varphi(ax + iy)\,d\mu(x) .$$

Then F_a is unitary if and only if $a = \sqrt{2}$. (cf. Gauss transform $\sigma = F_1$.)

<u>Observations.</u> Let $\{\xi_k\}$ be a c.o.n.s. for $L^2(R^1)$ and let $\varphi(x)$ be a tame function given by $\varphi(x) = f(< x,\xi_1 >,\ldots,< x,\xi_n>)$. Then $\varphi(x) \in (L^2)$ if and only if $f \in L^2(R^n, \exp(-\frac{1}{2}\sum_1^n t_j^2)\,dt^n)$.

The last condition is equivalent to

$$f(\sqrt{2}\,t_1,\ldots,\sqrt{2}\,t_n)\exp(-\tfrac{1}{2}\sum_1^n t_j^2) \in L^2(R^n, dt^n).$$

For such $\varphi(x)$, F formally acts in the following manner.

$$(F\varphi)(y) = \int f(\sqrt{2} < x,\xi_1> + i <y,\xi_1 >,\ldots,\sqrt{2} < x,\xi_n > + i <y,\xi_n>)$$

$$= (\sqrt{2\pi})^{-n} \int \ldots \int f(\sqrt{2}\,t_1+is_1,\ldots,\sqrt{2}\,t_n+is_n)\exp(-\tfrac{1}{2}\sum_1^n t$$

$$s_j = <y,\xi_j>,\ j = 1,2,\ldots$$

$$= (\sqrt{2\pi})^{-n} \int \ldots \int f(\sqrt{2}\,u_1,\ldots,\sqrt{2}\,u_n)\exp(-\tfrac{1}{2}\sum_1^n u_j^2)$$

$$\times \exp(i\sum_1^n u_j \frac{s_j}{\sqrt{2}}) \times \exp\left[\frac{1}{2}\sum_1^n \left(\frac{s_j}{\sqrt{2}}\right)^2\right] du^n$$

n short, form a function $f(\sqrt{2}\, t_1, \ldots, \sqrt{2}\, t_n)\exp(-\frac{1}{2}\sum_1^n t_j^2)$ which s guaranteed to be in $L^2(R^n, dt^n)$ and have the ordinary Fourier ransform denoted by $g(v_1, \ldots, v_n)$. Replacing v_j with $< y, \xi_j >/\sqrt{2}$ n the expression $g(v_1, \ldots, v_n)\exp[\frac{1}{2}\sum_1^n v_j^2]$ we obtain the Fourier-iener transform of φ.

We now state two simple properties of the Fourier-Wiener ransform.

°) Suppose $\varphi \in (L^2)$ is expressed in the form

$$\varphi(x) = \sum_{n \geq 0} \sum_k a_{n,k} \psi_{n,k}(x) \qquad \text{(Fourier series)} ,$$

here the $\psi_{n,k}$ are Fourier-Hermite polynomials of degree n. Then e have

$$(F\varphi)(y) = \sum_{n \geq 0} i^n \sum_k a_{n,k} \psi_{n,k}(y) .$$

°) Let U_g , $g \in O_\infty$, be given by

$$(U_g\varphi)(x) = \varphi(g^*x) , \; g^* \in O_\infty^* , \; \varphi \in (L^2) .$$

e unitary operator U_g commutes with F :

$$F(U_g\varphi) = U_g(F\varphi) \quad \text{for every} \quad \varphi \in (L^2) . \tag{3}$$

is relation is easily verified using the fact that μ is O_∞^* - invariant.

.2. Translations.

By the ordinary Fourier transform on $L^2(R^n)$ the translation s changed into multiplication by an exponential function. We expect imilar situation for (L^2) .

Fix $x_o \in E^*$. The translation $x \longrightarrow x+x_o$ determines a new measure μ_{x_o} on $(E^*, \textcircled{B})$:

(9) $$d\mu_{x_o}(x) = d\mu(x+x_o) .$$

It is natural to ask if μ_{x_o} is equivalent to μ. Let us begin with a naive observation. Let $\{\xi_n\}$ be a c.o.n.s. for $L^2(R^1)$. The strong law of large numbers tells us that

$$\lim_{n \to \infty} \frac{1}{N} \sum_{n=1}^{N} < x, \xi_n >^2 = 1 \quad \text{for almost all} \quad x.$$

On the other hand

$$\lim_{N \to \infty} \frac{1}{N} \Sigma < x+x_o, \xi_n >^2 = 1 + \lim_{N \to \infty} \frac{1}{N} (2 \sum_{1}^{N} < x_o, \xi_n > < x, \xi_n > + \sum_{1}^{N} < x$$

In order that the support, as it were, of μ_{x_o} coincide with that of μ the above limit has to be 1. If x_o is restricted to $L^2(R^1)$, the above requirement is satisfied. Even such a simple illustration shows a crucial difference between μ and ordinary Lebesgue measure which is of course translation invariant.

Now we assume that $x_o \in L^2(R^1)$. Note that the expression $< x, x_o >$ has meaning not as a continuous bilinear functional but as a member of (L^2). Let $C_{x_o}(\xi)$, $\xi \in E$, be the characteristic functional o

Then we have

$$C_{x_o}(\xi) = \int_{E^*} e^{i < x, \xi >} d\mu_{x_o}(x)$$

$$= \int_{E^*} e^{i< x-x_o, \xi >} d\mu_{x_o}(x-x_o)$$

$$= e^{-1< x_o, \xi >} \int_{E^*} e^{i< x, \xi >} d\mu(x)$$

$$= \exp\{- \tfrac{1}{2}\| \xi+ix_o\|^2 - \tfrac{1}{2}\|x_o\|^2\} .$$

($\| \ \|$ stands for the $L^2(R^1)$-norm)

$$= \int_{E^*} e^{i<x, \xi >} \cdot e^{-< x,x_o> - \frac{1}{2}\|x_o\|^2} d\mu(x) .$$

These formal computations lead us to state the following

<u>Theorem 9.2</u>. If x_o is in $L^2(R^1)$, then μ_{x_o} is equivalent to μ and the Radon-Nikodym derivative is given by

$$\frac{d\mu_{x_o}}{d\mu}(x) = \exp[- < x,x_o> - \tfrac{1}{2} \|x_o\|^2] . \tag{10}$$

Thus we can say that μ is <u>quasi-invariant</u> under the translations by $x_o \in L^2(R^1)$.

For detailed discussions about the above theorem we refer to Y. Umemura [27]. There it is shown that μ_{x_o} is equivalent to μ if and only if x_o is in $L^2(R^1)$.

We define an operator U_{x_o} by

$$(U_{x_o}\varphi)(x) = \varphi(x+x_o)\exp[- \tfrac{1}{2} < x,x_o> - \tfrac{1}{4} \|x_o\|^2] , \ \varphi \in (L^2) . \tag{11}$$

This operator plays the role of the translation: $f(x) \longrightarrow f(x+x_o)$

for $f \in L^2(R^1)$. The additional factor of an exponential function in the expression (11) is needed to make U_{x_o} unitary.

The relation between U_{x_o} and the Fourier-Wiener transform F is given by

Theorem 9.3.

$$F(U_{x_o} \varphi)(y) = \exp(-\tfrac{i}{2} < y, x_o >)(F\varphi)(y) , \quad \varphi \in (L^2) . \tag{12}$$

Proof.

$$F(U_{x_o}\varphi)(y) = \int_{E^*} \varphi(\sqrt{2}\, x + x_o + iy) \exp(-\tfrac{1}{2} < \sqrt{2}\, x + iy, x_o > - \tfrac{1}{4} \|x_o\|^2) d\mu$$

$$= \exp(-\tfrac{i}{2} < y, x_o > - \tfrac{1}{4} \|x_o\|^2) \int_{E^*} \varphi(\sqrt{2}\, x' + iy) \exp(-\tfrac{1}{2} < \sqrt{2} x'$$

$$+ \tfrac{1}{2} \|x_o\|^2) d\mu \left(x' + \frac{x_o}{\sqrt{2}} \right)$$

$$= \exp(-\tfrac{i}{2} < y, x_o >) \int \varphi(\sqrt{2}\, x' + iy) d\mu(x)$$

$$= \exp(-\tfrac{i}{2} < y, x_o >) . (F\varphi)(y) .$$

9.3. Infinite dimensional motion group $M_\infty(E)$.

Let $O_\infty^*(E)$ be the collection of duals g^* to $g \in O_\infty(E)$. The set $O_\infty^*(E)$ is a topological group with respect to the obvious group operation and topology. The group $O_\infty^*(E)$ is isomorphic to the group $O_\infty(E)$ by the correspondence: $O_\infty(E) \ni g^{-1} \leftrightarrow g^* \in O_\infty^*(E)$. Each member of $O_\infty^*(E)$ is a linear transformation of E^* and is a measure preserving transformation on (E^*, μ).

We now introduce the infinite dimensional motion group $M_\infty(E)$ which is given by

$$M_\infty(E) = \{m = (g^*, \xi);\ g^* \quad O_\infty^*(E), \xi \quad E\},$$

with the multiplication

$$m_1 \cdot m_2 = (g_1^*, \xi_1) \cdot (g_2^*, \xi_2) = (g_1^* g_2^*, \xi_1 + g_1^* \xi_2), \tag{13}$$

where $m_i = (g_i^*, \xi_i) \in M_\infty(E)$, $i = 1,2$. Let $m \in M_\infty(E)$ act on E^* in such a way that

$$mx = g^* x + \xi.$$

Then $M_\infty(E)$ acts in a similar manner to the finite dimensional motion group. The group $M_\infty(E)$ is also topologized in the usual way.

The operator U_m, $m \in M_\infty(E)$, on (L^2) is defined by

$$U_m = U_\xi U_g, \quad m = (g^*, \xi) \tag{14}$$

where

$$(U_\xi \phi)(x) = \exp\{-\tfrac{1}{2} < x, \xi > - \frac{\|\xi\|^2}{4}\}\ \phi(x + \xi) \quad \text{and}$$

$$(U_g \phi)(x) = \phi(g^* x), \qquad \phi \in (L^2).$$

Obviously each U_m is a unitary operator on (L^2) and the collection U_m, $m \in M_\infty(E)$, forms a group - that is, $\{U_m, m \in M_\infty(E); (\text{I}$ is a unitary representation of the group $M_\infty(E)$.

The following analysis of (L^2) and its connection with $M_\infty(E)$ is due to Orihara [29] and Kōno [28]. First we observe the transformation of U_m, $m \in M_\infty(E)$, by the Fourier-Wiener transform F. Set

$$V_m = FU_m F^{-1},$$

then we obtain

$$V_m = V_\xi V_g,$$

where

$$V_\xi = FU_\xi F^{-1}, \quad (V_\xi \phi)(x) = \exp\{-\frac{i}{2} < x, \xi >\}\phi(x) \tag{15}$$

and

$$V_g = FU_g F^{-1}, \quad (V_g \phi)(x) = \phi(g^* x). \tag{16}$$

The expression (15) is proved by using the formula (10), and (16) is the same as (8).

Remark 3. The multiple Wiener Integral $\mathcal{H}_n$ is not invariant under U_m, $m \in M_\infty(E)$, while $\mathcal{H}_n$ is invariant if m is restricted to $(g^*, 0)$, $g^* \in O^*_\infty(E)$. This leads us to think of the translation as an operator which changes the class $\mathcal{H}_n$.

Differential Operators. Consider one parameter groups of unitary operators

$$U_{t\xi} \text{ and } V_{t\xi}, \quad -\infty < t < \infty, \quad \xi \text{ fixed} \tag{17}$$

arising from the translation. They are continuous in t ; therefore, by Stone's theorem, there exist spectral representations:

$$U_{t\xi} = \int e^{it\lambda} dE_\xi(\lambda),$$

$$V_{t\xi} = \int e^{it\lambda} dF_\xi(\lambda),$$

where $\{E_\xi(\lambda)\}$ and $\{F_\xi(\lambda)\}$ are resolutions of the identity. The infinitesimal generators $A_\xi = \frac{d}{dt} U_{t\xi}\big|_{t=0}$ are expressed in the form

$$\frac{1}{i} A_\xi = \int \lambda dE_\xi(x), \quad \frac{1}{i} B_\xi = \int \lambda dF_\xi(\lambda) .$$

For a tame function $\phi(x)$ expressed in the form $\phi(x) = f(< x, \xi_1 >, \ldots, < x, \xi_n >)$ with an orthonormal system $\xi_1, \ldots, \xi_n$, we define

$$\frac{\partial}{\partial \xi_k} \phi(x) = \frac{\partial}{\partial t_k} f(t_1, \ldots, t_n) \Big|_{t_j = < x, \xi_j >, \ 1 \leq j \leq n} .$$

Any function in A' and M (for notation see §6.1.) can be expressed in the above form by choosing an othronormal system in which the given ξ is one of the members, and hence we can apply the operator $\frac{\partial}{\partial \xi}$.

Proposition 2.

i) $\mathcal{D}(A_\xi) \supset$ A', M,

$\mathcal{D}(B_\xi) \supset$ A', M.

ii) $A_\xi = \frac{\partial}{\partial \xi} - \frac{1}{2}\xi,$

$B_\xi = -\frac{i}{2}\xi,$ on A' and M,

where ξ is the multiplication operator given by $\xi \cdot \phi(x) = < x, \xi > \phi(x)$.

Proof. The assertion i) is obvious. For the proof of ii) we note that if ϕ is in A' or M the functions

$\phi(x+t\xi)\exp[-\frac{1}{2} < x,\xi > t - \frac{1}{4}\|\xi\|^2 t^2]$ and $\exp[-\frac{1}{2} < x,\xi > t]\phi(x)$ are differentiable with respect to the variable t in the sense of (L^2)-norm. Simple computations give us the expressions in ii).

Now we are ready to observe how the operators A_ξ and B_ξ act on the space $\mathcal{H}_n$. Let $\{\xi_n\}$ be a c.o.n.s. for $L^2(R^1)$, and consider the Fourier-Hermite polynomials of the form

$$\prod_k H_{n_k}\left(\frac{< x,\xi_k >}{\sqrt{2}}\right).$$

Then we have

$$A_{\xi_j}\prod_k H_{n_k}\left(\frac{< x,\xi_k >}{\sqrt{2}}\right) = \frac{d}{dt} H_{n_j}\left(\frac{< x,\xi_j > + t}{\sqrt{2}}\right)\Bigg|_{t=0} \times \prod_{k\neq j} H_{n_k}\left(\frac{< x,}{\sqrt{}}\right.$$

$$- \frac{1}{2} < x,\xi_j > \prod_k H_{n_k}\left(\frac{< x,\xi_k >}{\sqrt{2}}\right).$$

To obtain the explicit form of the result we divide into two cases:

Case 1. $n_j = 0$

$$A_{\xi_j}\prod_k H_{n_k}\left(\frac{< x,\xi_k >}{\sqrt{2}}\right) = -\frac{1}{\sqrt{2}}\prod_{k\neq j,\, n_j=1} H_{n_k}\left(\frac{< x,\xi_k >}{\sqrt{2}}\right)$$

Case 2. $n_j > 0$

$$A_{\xi_j}\prod_k H_{n_k}\left(\frac{< x,\xi_k >}{\sqrt{2}}\right) = \frac{1}{\sqrt{2}} n_j H_{n_j-1}\left(\frac{< x,\xi_j >}{\sqrt{2}}\right)\prod_{k\neq j} H_{n_k}\left(\frac{< x,\xi_k >}{\sqrt{2}}\right)$$

$$- \frac{1}{2\sqrt{2}} H_{n_j+1}\left(\frac{< x,\xi_j >}{\sqrt{2}}\right)\prod_{k\neq j} H_{n_k}\left(\frac{< x,\xi_k >}{\sqrt{2}}\right)$$

(c.f. formulas in Appendix)

In a similar manner we have

$$B_{\xi_j} \prod_k H_{n_k}\left(\frac{<x, \xi_k>}{\sqrt{2}}\right) = -\frac{i}{\sqrt{2}} \prod_{k \neq j, n_j = 1} H_{n_k}\left(\frac{<x, \xi_k>}{\sqrt{2}}\right), \text{ if } n_j = 0$$

$$= -\frac{i}{\sqrt{2}} n_j \; H_{n_j - 1}\left(\frac{<x, \xi_j>}{\sqrt{2}}\right) \prod_{n_k \neq n_j} H_{n_k}\left(\frac{<x, \xi_k>}{\sqrt{2}}\right)$$

$$- \frac{i}{2\sqrt{2}} H_{n_j + 1}\left(\frac{<x, \xi_j>}{\sqrt{2}}\right) \prod_{n_k \neq n_j} H_{n_k}\left(\frac{<x, \xi_k>}{\sqrt{2}}\right), \quad \text{if } n_j > 0.$$

Thus the operator A_ξ carries a function in $\mathcal{H}_n \cap \mathcal{D}(A_\xi)$ into $\mathcal{H}_{n-1} \oplus \mathcal{H}_{n+1}$. The same is true for B_ξ. (Note that $B_\xi = FA_\xi F^{-1}$ and that $\mathcal{H}_n$ is invariant under the operator F.)

Remark 3. For a physical interpretation of the above property we refer to H. Weyl [30, Chapter II].

9.4. Infinite dimensional Laplacian operator.

The infinite dimensional Laplacian operator has been discussed by P. Lévy [31, III partie] and Y. Umemura [32]. Following them we shall first introduce the finite dimensional operator Δ_n which will approximate the infinite dimensional Laplacian operator.

Let $\{\xi_n\}$ be a c.o.n.s. for $L^2(R^1)$, and let $\mathcal{B}_n$ be the smallest σ-field with respect to which the $<x, \xi_k>$, $k \leqq n$, are measurable. Being inspired by the observation on page 114 we define an operator Δ_n on $L^2(E^*, \mathcal{B}_n)$ by

$$\Delta_n = \sum_{k=1}^{n} \left(\frac{\partial^2}{\partial \xi_k^2} - \xi_k \cdot \frac{\partial}{\partial \xi_k}\right), \tag{18}$$

where $\frac{\partial}{\partial \xi}$ and $\xi \cdot$ stand for the same operators as in §9.3. The expression (18) looks like the Laplace-Bertrami operator on the sphere rather than the Laplacian on R^n. Suppose $\phi(x)$ is a Fourier-Hermite polynomial in $L^2(E^*, \mathcal{B}_n)$, say

$$\phi(x) = \prod_k H_{n_k}\left(\frac{\langle x, \xi_k \rangle}{\sqrt{2}}\right)$$

with $\Sigma\, n_k = p$. Then Δ_n can be applied to ϕ and we obtain

$$\Delta_n \phi = \sum_{k=1}^{n} \left(\frac{1}{2} H''_{n_k}\left(\frac{\langle x, \xi_k \rangle}{\sqrt{2}}\right) - \frac{\langle x, \xi_k \rangle}{\sqrt{2}} H'_{n_k}\left(\frac{\langle x, \xi_k \rangle}{\sqrt{2}}\right)\right) \prod_{j \neq k} H_{n_j}\left(\frac{\langle x, \xi_j \rangle}{\sqrt{2}}\right)$$

Using the differential equation for Hermite polynomials we finally obtain the following simple result:

$$\Delta_n \phi = -p\phi. \tag{19}$$

We now come to the definition of <u>the infinite dimensional Laplacian</u> Δ_∞ on (L^2):

$$\Delta_\infty = \lim_{n \to \infty} \Delta_n. \tag{20}$$

Although the expression (20) is formal, we can give a rigorous meaning as follows.

1) For any Fourier-Hermite polynomial ϕ based on the c.o.n.s. $\{\xi_n\}$ there exists n such that $\phi \in L^2(E^*, \mathfrak{B}_n)$. Since $\Delta_m \phi = -p\phi$ (p = the degree of ϕ) holds for every $m \geqq n$, $\Delta_\infty \phi$ is defined to be $-p\phi$. Obviously this definition does not depend on the choice of n.

2) The collection of all the Fourier-Hermite polynomials of degree p based on the c.o.n.s. $\{\xi_n\}$ is included in the domain of Δ_∞ and is dense in $\mathcal{H}_p$. The relation (19) shows that Δ_∞ is bounded if it is restricted to $\mathcal{H}_p$. Therefore Δ_∞ extends to a bounded linear operator on $\mathcal{H}_p$ (the extension is denoted by the same symbol Δ_∞). Obviously Δ_∞ is a symmetric operator. Thus each element of $\mathcal{H}_p$ is the eigenfunction of Δ_∞ with eigenvalue $-p$. As a consequence, we see that Δ_∞ does not depend on the choice of the ξ_n.

3) Δ_∞ can be defined on $\sum_p \mathcal{H}_p$ (algebraic sum) and is symmetric

(L^2).

Proposition 3. $\mathfrak{D}(\Delta_\infty) \supset A'$ and

$$\Delta_\infty e^{\langle x,\xi \rangle} = (\|\xi\|^2 - \langle x,\xi \rangle) e^{\langle x,\xi \rangle} \tag{21}$$

Proof. Choose a c.o.n.s $\{\xi_n\}$ with $\xi_1 = \xi/\| \ \|$. Then
ıple computation leads us to the formula (21). The first assertion
.lows immediately.

The following proposition is straightforward.

Proposition 4. The operator Δ_∞ commutes with the U_g, g O_∞.

Remark 4. For detailed discussions about Δ_∞ we refer to
Umemura [32]. It is interesting to note that he has given a
ıracterization of Δ_∞: if H is a symmetric operator on (L^2) with
ısonably rich domain (suffices to assume that $\mathfrak{D}(H) \supset A$) and if
is invariant under the group O_∞^*, then H is a function of Δ_∞.

Exercise. There are interesting relations between Δ_∞ and the
ırier-Wiener transform F. Further if Δ_∞ is discussed on $\mathcal{F}(C,E) = \tau\{(L^2)\}$,
have many intimate connection with P. Lévy's approach [31,III partie].
ıse discussions are left to the reader.

Bibliography

[25] R. H. Cameron and W. T. Martin, Fourier-Wiener transforms of functionals belonging to L_2 over the space C. Duke Math. J. 14 (1947), 99-107.

[26] N. Wiener, Hermitian polynomials and Fourier analysis. J. Math. Phys. (1939), 70-73.

[27] Y. Umemura, Measures on infinite dimensional vector spaces, Pub. of the Research Inst. for Math. Sci., Kyoto Univ. A. 1 (1965), 1-47.

[28] N. Kôno, Special functions connected with representations of the infinite dimensional motion group. J. of Math. Kyoto Univ. 6 (1966), 61-83.

[29] A. Orihara, Hermitian polynomials and infinite dimensional motion group, loc. cit. 1-12.

[30] H. Weyl, The theory of groups and quantum mechanics. Mathuen and Co. Ltd. 1931.

[31] P. Lévy, Problèmes concrets d'analyse fonctionelle. Gautier-Villars 1951.

[32] Y. Umemura, On the infinite dimensional Laplacian operator. J. Math. Kyoto Univ. 4 (1965), 477-492.

§10. Applications.

This section will be devoted to brief remarks on several applications of our analysis on the space (L^2).

10.1. N. Wiener's theory for nonlinear networks.

(On the fourth anniversary of N. Wiener's death)

We are going to discuss a stationary stochastic process obtained through a nonlinear network from a Brownian input. We refer to N. Wiener [19, Lecture 10] for a discussion of why a Brownian input is fitting for the analysis of networks. It is quite reasonable to assume that the given network is non-explosive, deadbeat, and so forth. The output through the network is a functional of the Brownian motion which is the input. To analyze the functional we prefer W.N. to Brownian motion. Thus our situation can be expressed by the following figure

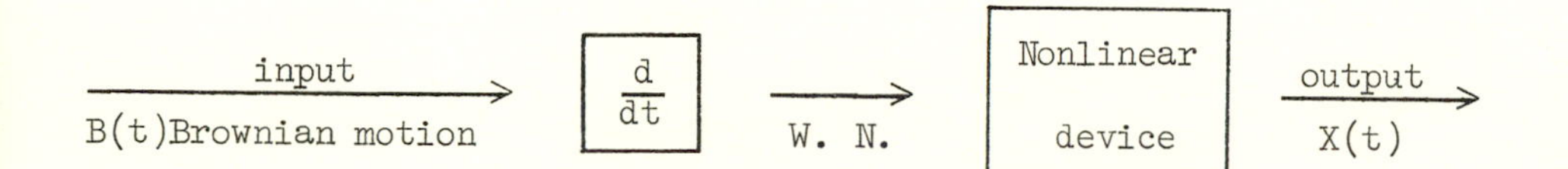

We can regard the output as a stationary stochastic process which shares the shift operator U_t with W. N. If the output $X(t)$ has a finite variance, we can express it in

the form

$$(1) \qquad X(t) = U_t\varphi(x), \qquad \varphi \in (L^2)$$

The output cannot be a functional of the future input, which means that $\varphi(x)$ is $\textcircled{B}_0$ -measurable, or equivalently $\varphi \in L^2(0)$ for W.N. (for notations see §4.5).

We are now able to appeal to our discussions in §6 and §7. Using the realization of the multiple Wiener integral, we obtain

$$(2) \qquad \varphi(x) = \sum_{n=0}^{\infty} \varphi_n(x), \qquad \varphi_n \in \mathcal{H}_n \cap L^2(0),$$

$$\varphi_n \sim F_n(t_1, \cdots, t_n), \quad F_n \in \widehat{L^2(R^n)}$$

with $\qquad F_n(t_1, \cdot\cdot, t_n) = 0$ for some $t_i < 0$, and

$$(3) \qquad U_t\varphi_n \sim F_n(t_1 - t, \cdots, t_n - t) = U_t^{(n)}F_n(t_1, \cdots, t_n).$$

The given network is therefore characterized by a system $\{F_n, n \leqq 0\}$ of functions. To determine the F_n we proceed as follows. We form networks through which $U_t\ \psi_n^m$, $\psi_n^m \in \mathcal{H}_n \cap L^2(0)$ are given corresponding to the W.N. input, where the spans $\mathfrak{S}\{U_t\ \psi_n^m, -\infty < + < \infty\}$, $m = 1,2,...$ are mutually orthogonal cyclic sub-spaces and their direct sum is the whole space

$\mathcal{H}_n$ (c.f. § 7. P.102). Indeed networks can be formed mechanically. Moreover we can choose the $\widehat{L^2}(R^n)$-function G_n^m so that they are normalized and

$$U_t^{(n)} F_n = \sum_m A_m U_t^{(n)} G_n^m, \quad -\infty < t < \infty,$$

holds. We then apply the ergodic theorem to obtain

$$A_m = \int_{E^*} \varphi_n(x)\psi_n^m(x) = \lim_{T\to\infty} \frac{1}{2T}\int_{-T}^{T} U_t\varphi_n(x)U_t\psi_n^m(x)dt, \quad \text{a.e.}(\mu),$$

which is equal to $\int \underset{R^n}{\cdots} \int F_n(t_1,\ldots,t_n)\, G_n^m(t_1,\ldots,t_n)dt_1 \ldots dt_n$.

Therefore, observing the outputs $U_t\varphi_n$ and $U_t\psi_n^m$ we can compute the coefficients A_n. Thus we have analyzed the given network, and the spectrum of the output $X(t)$ can easily be obtained.

10.2. Stochastic differential equations.

Consider a stationary process $X(t) = X(t,x)$, $x \in E^*(\mu)$ represented as an output for W.N. input:

$$X(t,x) = U_t\varphi(x), \qquad \varphi \in L^2(0).$$

Assume that $X(t)$ satisfies a stochastic differential equation of the following type:

$$(4) \qquad dX(t) = a(X(t))dt + b(X(t))\, dB(t), \qquad t > t_0 (\geqq -\infty)$$
$$X(t_0) = X_0$$

or equivalently

$$(4') \quad X(t) = X_0 + \int_{t_0}^{t} a(X(\tau))d\tau + \int_{t_0}^{t} b(X(\tau))dB(\tau)$$

The general theory of stochastic differential equation is found in K. Ito [33], Skorohod [34] and others. In this section we understand the equation (4') in the following manner. The first integral in the expression (4') should be understood as the Bochner integral, i.e. the integral is defined in a similar manner to the Riemann integral and the limit is taken in the sense of (L^2)-norm. Naturally we require that $a(X(t)) \in (L^2)$. Since $a(X(t)) = U_t\, a(X(0))$, it suffices to assume that $a(X(0)) \in (L^2)$. Moreover, under this assumption, we know that $a(X(t))$ is strongly continuous in t. The second integral should be a generalization of the Wiener integral (see §2). The increment $dB(t)$, $dt > 0$, of the Brownian input is independent of $b(X(\tau))$, $\tau \leq t$, therefore we expect that if $b(X(t)) \in \mathcal{H}_n$ then the product $b(X(t))dB(t)$ be in $\mathcal{H}_{n+1}$. These considerations lead us to define the integral based on Brownian motion. Suppose that $b(X(0))$ (nothe that it is in the subspace $L^2(0)$) has a realization $\{F_n(t_1,\cdots,t_n),\ n \geqq 0\}$, $F_n \in \widehat{L^2(R^n)}$, with the property that $F_n(t_1,\cdots,t_n) = 0$ if some $t_i > 0$. Since F_n is symmetric it suffices to look at F_n on the sector

$0 \geq t_1 \geq \cdots \geq t_n$. Define $b(X(t))dB(t)_{t=o}$ in such a way that it has a realization given by $\{\tilde{F}_n(t_1,\cdots,t_n),\ n \geq 0\}$ where

$$(5)\quad \tilde{F}_n(t_1,\cdots,t_n) = I_{dt}(t_1)F_{n-1}(t_2,\cdots,t_n),\ dt \geq t_1 \geq 0 \geq t_2 \geq \cdots \geq t_n.$$

I_{dt} being the indicator function of the interval $[0,dt]$. The function $\tilde{F}_n$ given by (5) extends to a $\widehat{L^2}(R^n)$-function. Thus we are given a realization of $B(X(t))dB(t)_{t=o}$. Since $b(X(\tau))dB(\tau)$ is obtained by applying the shift operator U_τ to $B(X(t))dB(t)_{t=o}$, we have a realization

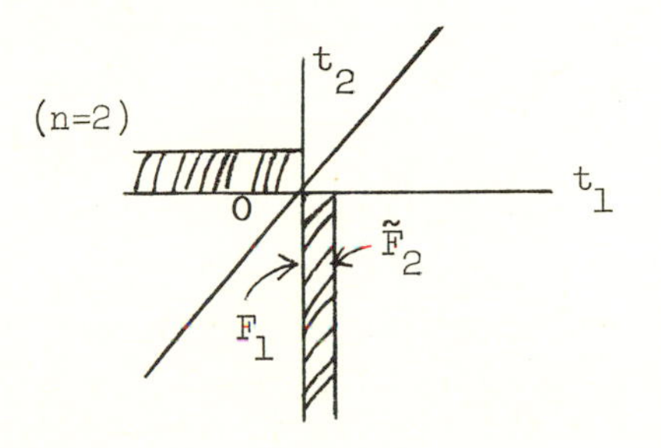

$$b(X(\tau))dB(\tau) \sim \{\tilde{F}_n(t_1-\tau,\cdots,t_n-\tau),\ n \geq 0\} \text{ with } \tilde{F}_o = 0.$$

Thus we can discuss the integrals, integrability and so on.

Sometimes the method presented here is useful to obtain the explicit expression of the solution for a certain kind of stochastic differential equations. Here are two examples.

Example 1.

$$(6)\qquad dX(t) = -\lambda X(t)dt + dB(t),\ -\infty < t < \infty,\quad \lambda > 0,\quad X(-\infty) = 0$$

(the so-called Langevin equation).

Suppose that $X(0)$ has the realization $\{F_n\}$. Then the left hand side of the equation (6) can be represented by the sequence of functions $d_t F_n(t_1 - t, \cdots t_n - t)$, $n \geq 0$, while the right hand side is represented by $-\lambda F_n(t_1 - t, \cdots, t_n - t)dt$, $n \neq 1$; $-\lambda F_1(t_1 - t) + I_{dt}(t_1 - t)$, $n=1$. Therefore we have a system of differential equations as follows.

$$d_t F_o = -\lambda F_o$$

$$d_t F_1(t_1 - t) = -\lambda F_1(t_1 - t) + I_{dt}(t_1 - t)$$

$$d_t F_n(t_1 - t, \cdots, t_n - t) = -\lambda F_n(t_1 - t, \cdots, t_n - t), \qquad n > 1.$$

By a simple computation we obtain the solution of the above equations only assuming that $F_n \in \widehat{L^2(R^n)}$, namely,

$$F_n = 0, \qquad n \neq 1$$

$$F_1(t_1) = \begin{cases} \exp[\lambda t_1], & t_1 \leq 0 \\ 0 \quad , & t_1 > 0. \end{cases}$$

Thus the process $X(t)$ is a stationary Gaussian process with covariance function $\frac{1}{2\lambda} \exp[-\lambda |t|]$.

<u>Example</u> 2. We consider a classical diffusion $X(t)$, $-\infty < t < \infty$, defined by

$$\text{(7)} \qquad dX(t) = (aX(t) + a')dt + (bX(t) + b')dB(t)$$

$$X(-\infty) = 0.$$

We are interested in finding an explicit expression of the solution for (7) assuming that $X(t)$ is expressed in the form $U_t\ \varphi(x)$, $\varphi \in (L^2) \cap L^2(0)$.

Suppose that $X(0) = \varphi(x)$ has the realization $\{F_n, n > 0\}$. By the same method as in Example 1 we obtain the following system of differential equations.

$$(8) \quad d_t F_n(t_1 - t, \cdots, t_n - t) = (aF_n(t_1 - t, \cdots, t_n - t) + a')dt +$$

$$+ (bF_{n-1}(t_2 - t, \cdots, t_n - t) + b'\ \delta_{n,1}) I_{dt}(t_1 - t),$$

$$, t \geq t_1 \geq t_2 \geq \cdots \geq t_n, \ n > 0, \ d_t F_o = aF_o\ dt + a' .$$

First we note that the stationarity and the initial condition $X(-\infty) = 0$ imply that $F_o = 0$, and hence a' must be 0. The equation (8) for $n = 1$ gives us the solution of the form

$$F_1(t_1) = \begin{cases} b' \exp[-at_1], & t_1 \leq 0 \\ 0, & t_1 > 0 . \end{cases}$$

We note that a must be negative in order for a solution to exist in $L^2(R^1)$. For the case $n = 2$ we come to a particular type of integral equation:

$$F_2(t_1 - t, t_2 - t) - F_2(t_1, t_2) = a \int_{\max(t_1, 0)}^{t} F_2(t_1 - \tau, t_2 - \tau) d\tau +$$

$$+ bb' \int_0^t e^{-a(t_2 - \tau)} I_{d\tau}(t_1 - \tau).$$

The solution of this is

$$F_2(t_1,t_2) = bb'\exp[-at_2], \qquad 0 \geq t_1 \geq t_2.$$

With this solution we can solve the equation (8) for the case $n = 3$, and so forth. Finally we obtain

$$(9) \qquad F_n(t_1,\cdots,t_n) = b^{n-1}\, b'\, \exp[-a \min_{1 \leq j \leq n} t_j], \qquad n \geq 1.$$

Thus we have determined the process completely.

Remarks for Example 2. By using the expression obtained above we can easily form the best (nonlinear) predictor for $X(t)$. Indeed we can prove that $\textcircled{B}_t(X) = \textcircled{B}\ (X(\tau), \tau \geq t)$ coincides with $\textcircled{B}_t$ for the W.N., i.e. for the $B'(t)$ arising in the equation (7). The best (in the sense of the least mean square error) predictor for $X(t)$, $t > 0$, under the assumption that $\{X(s),\ s \leq 0\}$ is observed is given by the conditional expectation $E(X(t)/\textcircled{B}_0(X))$ which is nothing but the projection of $X(t)$ on the subspace $L^2(0)$, where we use the relation $\textcircled{B}_0(X) = \textcircled{B}_0$. Furthermore, the best predictor has a realization $\{F_n(t_1-t,\cdots,t_n-t).\ I(t_1,\cdots,t_n),\ n \geq 0\}$, where I is the indicator function of the sector $0 \geq t_1 \geq t_2 \geq \cdots \geq t_n$. (cf M. Nisio [35])

10.3. Canonical commutation relation

We consider the connonical commutation relation arising in quantum mechanics. For a system of finite, say n, degree of freedom we have the following relations between momentum operators p_j, $1 \leq j \leq n$, and position operators q_j, $1 \leq j \leq n$:

$$(10) \qquad [q_j p_k] = i\hbar\delta_{jk}, \quad [p_j, p_k] = [q_j, q_k] = 0.$$

The operators p_j and q_j are represented on the Hilbert space $L^2(R^n)$ in such a way that

$$(11) \qquad p_j\varphi(x) = \hbar \frac{1}{i} \frac{\partial}{\partial x_j} \varphi(x), \qquad x = (x_1, \cdots, x_n),$$
$$q_j\varphi(x) = x_j\varphi(x), \qquad \varphi \in L^2(R^n).$$

Moreover it is known that the irreducible representation of the p_j's and the q_j's satisfying (10) is unique up to unitary equivalence. (see von Neumann [36]).

We expect a similar situation in the case of a Boson field. There are given self-adjoint operators p_λ, q_λ, λ real which satisfy the following (formal) relations.

(12) $$[q_\lambda, p_{\lambda'}] = i\hbar\delta(\lambda-\lambda'), \quad (\delta \text{ is the } \delta\text{-function}),$$

$$[p_\lambda, p_{\lambda'}] = [q_\lambda, q_{\lambda'}] = 0.$$

Here the p_λ and the q_λ must be thought of as operator densities rather than operators. Therefore it is fitting to introduce the smeared operators

$$p(\xi) = \int \xi(\lambda)p_\lambda d\lambda \quad \text{and} \quad q(\xi) = \int \xi(\lambda)q_\lambda d\lambda, \ \xi \in E.$$

Then the commutation relation (12) becomes

(12') $$[p(\xi), q(\eta)] = i\hbar < \xi, \eta > ,$$

$$[p(\xi), p(\eta)] = [q(\xi), q(\eta)] = 0.$$

Form unitary operators $P(\xi) = \exp[ip(\xi)]$ and $Q(\xi) = \exp[iq(\xi)]$, and set $\hbar = 1$. Then we are given the relations

(12") $$P(\xi)Q(\eta) = \exp(i < \xi, \eta >)$$

$$P(\xi)P(\eta) = P(\xi+\eta) = P(\eta)P(\xi)$$

$$Q(\xi)Q(\eta) = Q(\xi+\eta) = Q(\eta)Q(\xi).$$

With these set-up we now follow the approach due to Y.

Umemura ([27] and others). We are interested in finding operators $P(\xi)$, $Q(\xi)$, $\xi \in E$, satisfying as many as possible of the conditions:

i) $P(\xi)$, $Q(\xi)$, $\xi \in E$, are unitary operators acting on a Hilbert space,

ii) the mappings $E \ni \xi \longrightarrow P(\xi)$, $\xi \longrightarrow Q(\xi)$ are continuous

iii) the relations (12") hold

iv) irreducibility.

A representation of the $P(\xi)$ and the $Q(\xi)$ is given in the following manner. Take the W.N. and consider the space (L^2). We define operators $P(\xi)$ and $Q(\eta)$ by

$$(13) \qquad P(\xi)\varphi(x) = \varphi(x+\xi)\sqrt{d\mu_\xi/d\mu(x)}$$

$$Q(\eta)\varphi(x) = \exp[i < x, \eta >]\ \varphi(x), \qquad \varphi \in (L^2)$$

It is quite easy to see that the operators $P(\xi)$, $Q(\xi)$, $\xi \in E$, satisfy the above conditions i) ii) and iii). It is interesting to note that the operator $P(\xi)$ given by (13) is somewhat different from the operator $\exp[ip_j]$, where p_j is given by (11). This difference comes from the property that the measure

μ of W.N. is not invariant under the translation but quasi-invariant (see §9.2.). However the relation $F\, P(\xi)F^{-1} = Q(-\frac{1}{2}\xi)$ (F is the Fourier-Wiener transform) similar to the relation arising from p_j and q_j on $L^2(R^n)$ holds even in the case of $P(\xi)$ and $Q(\xi)$ on (L^2). (see also §9.2.).

Furthermore Y. Umemura has obtained a necessary and sufficient condition under which two representations of the $P(\xi)$ and the $Q(\xi)$ are equivalent. As a result we can form uncountably many inequivalent representations. This is a striking result which is quite different from the case of finite degree of freedom. As for the irreducibility he also has discovered a criterion in terms of the ergodic property of measures from which the representation spaces (L^2-spaces) are formed.

[Bibliography]

[33] K. Itô, On stochastic differential equations, 1951, Amer. Math. Soc.

[34] A. V. Skorohod, Studies in the theory of random processes (Russian) 1961, English translation, Addison-Wesley 1965.

[35] M. Nisio, Remarks on the canonical representation of strictly stationary processes, J. Math. Kyoto Univ. 1(1961) 129-146.

[36] J. von Neumann, Die Eindeutigkeit der Schrödingerschen Operatoren, Math. Ann. 104 (1931) 570-578.

§.11. Generalized White Noise.

Up to now we have been concerned with Gaussian white noise, which was, roughly speaking, the derivative of Brownian motion. We shall generalize it to the derivative of a Lévy process with stationary increments, that is, a generalized white noise.

Let E be a nuclear space such that

$$E \subset L^2(R^1) \subset E^*. \tag{1}$$

The generalized white noise is a stochastic process the characteristic functional of which is given by

$$\begin{aligned} C(\xi) = \exp[im\langle \xi, 1 \rangle \; + \\ + \sigma^2\|\xi\|^2/2 + \int_{-\infty}^{\infty}\int_{-\infty}^{\infty} (e^{i\xi(t)u} - 1 - \frac{i\xi(t)u}{1+u^2}) dn(u)dt], \quad \xi \in E. \end{aligned} \tag{2}$$

In particular, if $C(\xi)$ is expressed in the form

$$\exp[\int_{-\infty}^{\infty}(e^{i\xi(t)u} - 1 - \frac{i\xi(t)u}{1+u^2})dt], \tag{3}$$

the corresponding process is called the Poisson white noise with jump u.

Observing the expression (2) we understand that a generalized white noise is composed of three kinds of basic processes, namely constant process, Gaussian white noise and Poisson white noises with various jumps. We are therefore interested in the investigation of Poisson white noise which is another basic process different from Gaussian white noise. We shall further discuss stationary processes obtained by an integral of Poisson white noises with the jump u with respect to the Lévy measure $dn(u)$ (cf. § 3.3).

11.1. Poisson white noise. Given a Poisson white noise with jump 1, we compensate the constant term to obtain a characteristic functional

$$(4) \qquad C_P(\xi) = \exp[\int_{-\infty}^{\infty}(e^{i\xi(t)} - 1 - i\xi(t))dt], \qquad \xi \in E,$$

and the associated measure space (E^*, μ_P) such that

$$C_P(\xi) = \int_{E^*} e^{i\langle x, \xi \rangle} d\mu_P(x).$$

This is also called Poisson white noise. Obviously it is a stationary process: $C_P(S_t\xi) = C_P(\xi)$, and further it satisfies the following properties:

i) it has independent value at every moment,

ii) $\int_{E^*} |\langle x, \xi \rangle|^n d\mu_P(x) < \infty$ for every n,

iii) $\int_{E^*} \langle x , \xi \rangle d\mu_P(x) = 0.$

Besides the above three properties, the Poisson white noise has many properties similar to those of Gaussian white noise. We shall therefore confine ourselves here to interesting differences.

We first observe the Hilbert space $L^2(E^*, \mu_P)$. The polynomial and the exponential function can be defined in a similar manner to the case of Gaussian white noise, and they are dense in $L^2(E^*,\mu_P)$. The linear functional $\langle x , \xi \rangle, \xi \in E$, extends to $\langle x, f \rangle, f \in L^2(R^1)$, in the space $L^2(E^*, \mu_P)$. This situation is quite the same as in § 7. An interesting difference appears in a realization of $L^2(E^*, \mu_P)$ using the transformation τ defined in § 4.4:

$$(\tau\varphi)(\xi) = \int_{E^*} e^{i\langle x, \xi \rangle} \varphi(x)d\mu_P(x), \qquad \varphi \in L^2(E^*, \mu_P).$$

For example we have

$$\tau(e^{i\langle x, \eta\rangle})(\xi) = C_P(\xi+\eta) = \exp[\int_{-\infty}^{\infty} (e^{i(\xi(t)+\eta(t))} -1 -i(\xi(t) + \eta(t))dt]$$

$$\tau(\langle x , \eta \rangle)(\xi) = \tau[(\frac{1}{i} \frac{d}{dt} e^{i\langle x,\eta \rangle})|_{t = 0}]$$

$$= C_P(\xi)\int_{-\infty}^{\infty} \eta(t)(e^{i\xi(t)} -1)dt.$$

Set $P(x) = e^{ix} - 1$. If η_i, $1 \leqq i \leqq p$, have disjoint supports, then we have

$$(5) \quad \tau\left(\prod_{j=1}^{p} \langle x, \eta_j \rangle\right)(\xi) = C_P(\xi) \prod_{j=1}^{p} \int_{-\infty}^{\infty} \eta_j(t) P(\xi(t))dt, \qquad \xi \in E,$$

$$= C_P(\xi) \int \underset{R^p}{\cdots} \int F(t_1, \ldots, t_p) \prod_{j=1}^{p} P(\xi(t_j))dt^p,$$

where $F(t_1, \ldots, t_p)$ is the symmetrization of the product $\eta(t_1)\ldots\eta(t_p)$.

We now define

$$(6) \quad \mathcal{F}_p = \{f;\ f(\xi) = C_P(\xi) \int \underset{R^p}{\cdots} \int F_p(t_1, \ldots, t_p) P(\xi(t_1)) \ldots P(\xi(t_p)) \ldots$$

$$F_p \in \widehat{L^2}(R^p)\}$$

to obtain the following theorem:

<u>Theorem</u> 11.1. The Hilbert space $\mathcal{F} = \tau(L^2(E^*, \mu_P))$ has the direct sum decomposition of the form

$$(7) \quad \mathcal{F} = \sum_{p=0}^{\infty} \oplus\ \mathcal{F}_p,$$

and it holds that, for $f, g \in \mathcal{F}_p$.

$$(8) \quad (f, g)_{\mathcal{F}} = p\,!\ (F_p, G_p)_{L^2(R^p)},$$

where F_p and G_p are the $\widehat{L^2(R^p)}$-functions arising in the expression (6) corresponding to f and g, respectively.

<u>Idea of the proof</u>. In order to prove the theorem we proceed the following steps. Recall that the functional $<x,\xi>$, $\xi \in E$, extends to $<x, f>$, $f \in L^2(R^1)$, in the Hilbert space $L^2(E^*, \mu_P)$. In particular $<x, I_E>$, I_E being thé indicator function of the set E, is well defined and it is a r.v. on (E^*, μ_P), the distribution of which is the centered Poisson process:

$$\mu(\{x; <x, I_E> = k - \lambda\}) = \frac{\lambda^k}{k} e^{-\lambda}, \quad k = 0, 1, 2, \ldots$$

Where $\lambda = |E|$ the Lebesgue measure of E.

Then we can prove that the collection of all the polynomials of the $<x, I_E>$, E Borel set of finite Lebesgue measure, spanns the entire space $L^2(E^*, \mu_P)$. The subspace $\mathcal{F}_p$ of $\mathcal{F} = \tau(L^2(E^*, \mu_P))$ arising in the direct sum decomposition (6) corresponds to the subspace $\mathcal{H}_p$ of $L^2(E^*, \mu_P)$ which is spanned by the product $\prod_{j=1}^{p} <x, I_{E_j}>$, E_j disjoint.

To complete the proof we can appeal to the expression (5) replacing η_j with I_{E_j}.

11.2. Poisson-Charlier polynomials.

For Poisson white noise the so-called Poisson Charlier polynomial plays the role of the Hermite polynomial for Gaussian white noise in the analysis of the $(L)^2$-space.

Let $p(x,\lambda)$ be the Poisson distribution with parameter $\lambda > 0$:

$$(9) \qquad p(x;\lambda) = \frac{\lambda^x}{x} e^{-\lambda}, \quad x = 0, 1, 2, \ldots .$$

The Poisson-Charlier polynomial $p_n(x, \lambda)$ is given by

$$(10) \qquad p_n(x;\lambda) = \frac{\lambda^{\frac{n}{2}}}{n} (-1)^n \frac{\Delta^n p(x;\lambda)}{p(x;\lambda)},$$

where Δ is the difference operator: $\Delta f(x) = f(x) - f(x-1)$. The polynomials $p_n(x;\lambda)$, $n = 0, 1, 2, \ldots$, form an orthonormal system in the following sense:

$$(11) \qquad \sum_{x=0}^{\infty} p_n(x;\lambda)p_m(x;\lambda)p(x;\lambda) = \delta_{n,m}.$$

The generating function of the $p_n(x;\lambda)$ is

$$G(x,w;\lambda) \equiv \sum_{n=0}^{\infty} \frac{1}{\sqrt{\lambda^n}\, n!} p_n(x;\lambda)w^n = e^{-w}(1 + \frac{w}{\lambda})^x. \tag{12}$$

We now come to a discussion about the functional of the form

$$q_n(\langle x, I_E\rangle; \lambda) = p_n(\langle x, I_E\rangle + \lambda; \lambda), \quad \lambda = |E|.$$

Consider the generating function

$$\sum_{n=0}^{\infty} \frac{1}{\sqrt{\lambda^n}\, n!} q_n(\langle x, I_E\rangle;\lambda)w^n = e^{-w}(1 + \frac{w}{\lambda})^{\langle x, I_E\rangle + \lambda}$$

and apply the transformation τ. Then the right hand side turns out to be the following:

$$\tau(e^{-w}(1 + \frac{w}{\lambda})^{\langle x, I_E\rangle + \lambda})$$

$$= e^{-w} \int \exp[i\langle x, \xi\rangle + (\langle x, I_E\rangle + \lambda)\log(1 + \frac{w}{\lambda})]d\mu(x)$$

$$= \exp[-w + \lambda \log(1 + \frac{w}{\lambda}) + \int e^{i\xi(t) + I_E(t)\log(1 + \frac{w}{\lambda})} - 1 - i\xi(t) -$$

$$- I_E(t)\log(1 + \frac{w}{\lambda})]$$

$$= C_P(\xi)\exp[-w + \lambda \log(1 + \frac{w}{\lambda}) + \int_E(e^{i\xi(t) + I_E(t)\log(1 + \frac{w}{\lambda})} -$$

$$- e^{i\xi(t)} - \log(1 + \frac{w}{\lambda}))dt]$$

$$= C_P(\xi)\exp[w \int_E P(\xi(t))dt], \quad P(x) = e^{ix} - 1.$$

Therefore we obtain

$$\tau\{q_n(<x, I_E>; \lambda)\} = C_P(\xi) \frac{1}{\sqrt{\lambda^n \; n!}} (\int I_E(t) P(\xi(t)) dt)^n.$$

In a similar manner we can discuss the product of the functionals

$$\sum_{\ell=0}^{\infty} \frac{q_\ell(<x, I_E>; \lambda)}{\sqrt{\lambda^\ell \; \ell!}} u^\ell \sum_{m=0}^{\infty} \frac{q_m(<x, I_F>; \mu)}{\sqrt{\mu^m \; m!}} v^m \cdots \sum_{n=0}^{\infty} \frac{q_n(<x, I_G>; \nu)}{\sqrt{\nu^n \; n!}} w^n$$

$$= e^{-(u+v+\cdots+w)} (1 + \frac{u}{\lambda})^{<x, I_E> - \lambda} (1 + \frac{v}{\mu})^{<x, I_F> - \mu} \cdots (1 + \frac{w}{\nu})^{<x, I_G> -}$$

where the sets E, F,..., G are disjoint and $\lambda, \mu, \ldots, \nu$ denote the Lebesgue measure of E, F,..., G, respectively. Noting that the r.v.'s $<x, I_E>, <x, I_F>, \ldots, <x, I_G>$ are mutually independent, we apply the transformation τ to both sides in the above expression. Then we have

$$(13) \quad \tau\{q_\ell(<x, I_E>; \lambda) q_m(<x, I_F>; \mu) \cdots q_n(<x, I_G>; \nu)\}$$

$$= C_P(\xi) \frac{1}{\sqrt{\lambda^\ell \; \mu^m \cdots \nu^n} \sqrt{\ell! \; m! \cdots n!}} (\int I_E(t) P(\xi(t)) dt)^\ell$$

$$(\int I_F(t) P(\xi(t)) dt)^m \cdots (\int I_G(t) P(\xi(t)) dt)^n.$$

In other words, for a functional of the form

$q_\ell(<x, I_E> ; \lambda) \cdot q_m(<x, I_F> ; \mu) \cdots q_n(<x, I_G> ; \nu)$ we can associate an $L^2(R^{\ell+m+\cdots+n})$-function

$$\frac{1}{\cdots \sqrt{\nu^n} \sqrt{\ell!\, m! \cdots n!}} \cdot I_E(t_1) \cdots I_E(t_\ell) I_F(t_{\ell+1}) \cdots I_F(t_{\ell+m}) \cdots I(t_{\ell+m+\cdots+n}).$$

This function may be symmetrized so that we are given an illustration to Theorem 11.1. It should be noted that in the above correspondence by τ the relation (8) holds. In fact the product of the q_n has $L^2(E^*, \mu)$-norm 1, while the symmetrization of the above $L^2(R^{\ell+m+\cdots+n})$-function has norm $\frac{1}{\sqrt{(\ell+m+\cdots+n)!}}$ as is easily seen.

<u>Remark</u>. The product of the q_n plays not exactly the same role as the Fourier-Hermite polynomial for Gaussian white noise. For one thing, we can not form a <u>complete</u> orthonormal system for $L^2(E^*, \mu_P)$ by forming products of the q_n, but we are given orthonormal system by them.

11.3. <u>Stable white noise</u>.

Let μ_P be the measure of Poisson white noise introduced in the space E^*. We see that the group G^* of μ_P-measure preserving linear isomorphisms of E^* is very poor compared with the group $O_\infty^* = \{g^*; g \in O_\infty\}$ defined in connection with the Gaussian white noise. In fact, we do not know any reasonable one-parameter subgroup of G^* other than the shift.

Having been inspired by the discussion in § 3.4., we shall sum up Poisson white noises with various jumps so that the sum has certain invariant properties under some transformations acting on the space E^*. Recall that the characteristic functional of Poisson white noise with jump u is given by

$$C_{P,u}(\xi) = \exp\{\int(e^{i\xi(t)u} - 1 - i\xi(t)\, u)\, dt\}$$

The transformation g_a of the form

$$(14) \qquad E \ni \xi \to (g_a\xi)(t) = \xi(at)b, \quad b = b(a), \quad a > 0,$$

defines the adjoint g_a^* on E^*, by which we are given a measure μ_P' on E^* in such a way that

$$\mu_P' = g_a^* \circ \mu_{P,u}.$$

The characteristic functional of μ_P' is, therefore, $C_{P,u}(g_a\xi)$, which turns out to be

$$\exp\{\frac{1}{a}\int(e^{i\xi(t)bu} - 1 - i\xi(t)bu)dt)\}.$$

Thus in order to obtain a generalized white noise which is invariant under the g_a, $\underline{a}$ real positive, we consider a system

$\{\mu_{P,u}$; u real$\}$ and form a probability measure with the characteristic functional

$$C(\xi) = \exp\{\iint (e^{i\xi(t)u} - 1 - i\xi(t)u)dn(u)dt\},$$

where the Lévy measure dn is supposed to satisfy

$$\frac{1}{a}\, dn(\frac{u}{b}) = dn(u) \tag{15}$$

with some function $b(a)$.

<u>Lemma</u>. Suppose that a measure $dn(u)$ on R^1 satisfies the conditions:

i) $\int \frac{u^2}{1 + u^2}\, dn(u) < \infty$;

ii) With every a we can associate a constant b (depending on a) so that the equality (15) holds.

Then there exists α with $0 < \alpha < 2$ such that $dn(u)$ is of the form

$$C_{-}|u|^{-(a+1)}du, \quad \text{on} \quad (-\infty, 0)$$

and

$$C_{+}u^{-(a+1)}du, \quad \text{on} \quad (0, \infty),$$

where C_- and C_+ are nonnegative constants. The function b of a is necessarily expressed as $b(a) = a^{\alpha}$, $a > 0$.

With this particular choice of $b(a)$ in the expression (14) we now define an operator $g_a^{(\alpha)}$ by

$$(16) \qquad (g_a^{(\alpha)}\xi)(t) = \xi(at)a^{1/\alpha}.$$

Then, the following assertion follows immediately.

<u>Proposition</u>. Suppose that $C(-\xi) = C(\xi)$ and that the measure μ on E^* given by a generalized white noise is invariant under the $g_a^{(\alpha)*}$ the adjoint of $g_a^{(\alpha)}$ defined by (16) for every $a > 0$. Then the characteristic functional $C(\xi)$ of μ is expressed in the form

$$(17) \quad \exp\{\iint(\cdots)u^{-(\alpha + 1)}dudt\} \to \exp\{\iint(\cdots)|u|^{-(\alpha + 1)}dudt\}$$

This formula can be written in the form $\exp\{\text{const.} \int|\xi(t)|^{\alpha}dt\}$ and is denoted by $C_{\alpha}(\xi)$ (see § 3.4.). The measure with the characteristic functional $C_{\alpha}(\xi)$ given by (17) will be denoted by μ_{α}, and the stationary process $\mathbb{P}_{\alpha} = (E^*, \mu_{\alpha}, \{T_t\})$ is called a stable white noise with characteristic exponent α.

We now come to an investigation of the projective invariance for a stable white noise with characteristic exponent α. Let E_α be the nuclear space defined by

$$E_\alpha = \{\xi(u);\ \xi \in \mathbb{C}^\infty,\ \xi(\tfrac{1}{u})|u|^{\alpha/2} \in \mathbb{C}^\infty\}$$

with the natural topology similar to that introduced in the space D_0 appeared in § 8.1. With this choice of a nuclear space E_α, we can prove that each $g_a^{(\alpha)}$ is a linear isomorphism of E_α and that

$$\int |(g_a^{(\alpha)}\xi)(t)|^\alpha \, dt = \int |\xi(t)|^\alpha \, dt.$$

This equality yields the relation

$$C_\alpha(g_a^{(\alpha)}\xi) = C_\alpha(\xi),$$

which proves that

$$g_a^{(\alpha)*} \circ \mu_\alpha = \mu_\alpha. \tag{18}$$

Let us introduce the group $G(\mathcal{P}_\alpha)$ associated with the stable white noise $\mathcal{P}_\alpha$: It is the collection of all linear transformations g of E_α onto itself satisfying

i) g is a homeomorphism of E_α,

ii) $\int |g\xi(t)|^\alpha dt = \int |\xi(t)|^\alpha dt$.

Obviously the measure μ_α is invariant under the transformation

g^* the adjoint of $g \in G(\mathbb{P}_\alpha)$.

The collection $\{g_a^{(\alpha)}; a > 0\}$ forms a subgroup of $G(\mathbb{P}_\alpha)$, however we shall be able to present a much wider wubgroup H of $G(\mathbb{P}_\alpha)$ which is isomorphic to the projective linear group $PLG(2, \mathbb{R})$. For each element $\bar{h} \in PLG(2, \mathbb{R})$, represented in the matrix form $\bar{h} = \begin{pmatrix} a & b \\ c & d \end{pmatrix}$, there corresponds a transformation h of E_α given by

$$(19) \qquad \bar{h} \to (h\xi)(u) = \xi\left(\frac{au + b}{cu + d}\right)|cu + d|^{-2/\alpha}.$$

It is easy to see that such an h is a member of the group $G(\mathbb{P}_\alpha)$. We denote the group $\{h; \bar{h} \in PLG(2,\mathbb{R})\}$ by H. Then we have

<u>Theorem</u> 11.2. The group $G(\mathbb{P}_\alpha)$ admits a subgroup H which is isomorphic to $PLG(2, \mathbb{R})$ by the correspondence (19).

The theorem may be said to be a rephrase of the projective invariance of a symmetric stable process. (See T. Hida [37]). We can illustrate this fact in the following manner. To be somewhat more specific, let us consider the case $1 < \alpha < 2$. As in the case of Gaussian white noise, the function $< x, \xi >$, $\xi \in E_\alpha$, of x extends to a random variable $< x, I_{[0,t]}>$ on (E_α^*, μ_α), where $I_{[0,t]}$ is the indicator function of the interval $[0,t]$. Observing the characteristic function of $< x, I_{[0,t]}>$ we see that $X(t, x) = \langle x, I_{[0,t]} \rangle$, $t \geqq 0$, is a symmetric stable process with characteristic exponent α. Now we apply a transformation h^* with $h \in H$. Then $X(t, h^*x)$, $t \geqq 0$, is the same (stable) process as $X(t, x)$, $t \geqq 0$. By the definition of $X(t, x)$, h^* behaves as a transformation of the time variable t of $X(t, x)$ with a multiplicative constant. The reader will easily find a similarity with the discussion given in § 8.4. for Lévy's projective invariance of Brownian motion.

[Bibliography]

[37] T. Hida, Sur l'invariance projective pour les processus symétriques stables. C. R. Acad. Sc. Paris t. 267 (1968), 821-823.

[38] T. Hida, I. Kubo, H. Nomoto and H. Yosizawa, On projective invariance of Brownian motion. Pub. Research Inst. for Math. Sci. Kyoto Univ. vol. 4 (1969), 595-609.

[Appendix]

Definition of the Hermite Polynomial: $H_n(x) = (-1)^n e^{x^2} \frac{d^n}{dx^n} e^{-x^2}$

(1) $H_n''(x) - 2xH_n'(x) + 2nH_n(x) = 0$

(2) $H_n'(x) = 2nH_{n-1}(x)$

(3) $H_{n+1}(x) - 2xH_n(x) + 2nH_{n-1}(x) = 0$

(4) Generating Function $\sum_0^\infty \frac{t^n}{n!} H_n(x) = e^{-t^2+2tx}$

(5) $H_n(xy + \sqrt{1 - x^2}\, t) = \sum_0^n \binom{n}{k} x^{n-k}(1 - x^2)^{\frac{k}{2}} H_{n-k}(y)H_k(t)$

(5') $H_n\left(\frac{(a, x)}{\|a\|}\right) = \frac{1}{\|a\|^n} \sum_{k_1+\cdots+k_\ell=n} \frac{n!}{k_1! \cdots k_\ell!} \prod_{j=1}^{\ell} a_j^{k_j} H_{k_j}(x_j)$, $a = (a_1, \cdots, a_\ell)$, $x = (x_1, \cdots, x_\ell)$

(6) $\int H_n(xy + \sqrt{1 - x^2}\, t)H_k(t)e^{-t^2} dt = \frac{\sqrt{\pi}\, 2^k n!}{(n-k)!} x^{n-k}(1-x^2)^{\frac{k}{2}} H_{n-k}(y)$

(7) $\int H_n(x)H_m(x)e^{-x^2} dx = 2^n n! \sqrt{\pi}\, \delta_{n,m}$

$$\left\{\frac{1}{\sqrt{2^n n!}\sqrt[4]{\pi}} H_n(x)e^{-\frac{x^2}{2}}, n \geq 0\right\} \text{ c.o.n.s. in } L^2(\mathbb{R}^1)$$

(8) $\int H_\ell(x)H_m(x)H_n(x)e^{-\frac{x^2}{2}}\,dx = \dfrac{2^{\frac{\ell+m+n}{2}}\sqrt{\pi}\,\ell!\,m!\,n!}{(g-\ell)!\,(g-m)!\,(g-n)!}$, $g = \dfrac{\ell+m+n}{2}$

(9) $\sum_0^\infty \dfrac{x^n H_n(y)H_n(z)}{2^n n!} = (1 - x^2)^{-\frac{1}{2}} \exp\left[\dfrac{2xyz - (y^2 - z^2)x^2}{1 - x^2}\right]$

(10) $\sum_0^\infty \dfrac{x^n H_n(y)H_{n+k}(z)}{2^n n!} = (1-x^2)^{-\frac{k+1}{2}} \exp\left[\dfrac{2xyz - (y^2 + z^2)x^2}{1 - x^2}\right] H_k\left(\dfrac{z-xy}{\sqrt{1-x^2}}\right)$

(11) $\sum_0^n \dfrac{H_\ell(x)H_\ell(y)}{2^\ell \ell!} = \dfrac{H_{n+1}(x)H_n(y) - H_n(x)H_{n+1}(y)}{(2^{n+1}n!)(x - y)}$

(12) $\sum_{\ell=0}^n \dfrac{H_\ell(x)^2}{2^\ell \ell!} = \dfrac{1}{2^n n!}\left[(n+1)H_n(x)^2 - nH_{n+1}(x)H_{n-1}(x)\right]$

(12') $H_n(x)^2 \geqq \dfrac{n}{n + 1} H_{n+1}(x)H_{n-1}(x)$

<u>Remark</u>. Formulas (11) and (12) are used to prove that the c.o.n.s. $\{(\sqrt{2^n n!}\ \sqrt[4]{\pi}\,)^{-1} H_n(x)e^{-\frac{x^2}{2}}\}$ is "normalment dense".

(13) $\int e^{ixy}H_n(x)e^{-\frac{x^2}{2}}\,dx = \sqrt{2\pi}\; i^n e^{-\frac{y^2}{2}} H_n(y)$

(14) $\int e^{ixy}H_n(x)e^{-x^2}\,dx = \sqrt{\pi}\,(iy)^n e^{-\frac{y^2}{4}}$

(15) $\int e^{ixy}y^n e^{-y^2}\,dy = \left(\frac{i}{2}\right)^n \sqrt{\pi}\; H_n(x)e^{-\frac{x^2}{4}}$

16) $$\sum_{n=0}^{\infty} \frac{(iy)^n}{2^n n!} H_n(x) = e^{-\frac{y^2}{4} + ixy}$$

17) $$\sum_{n=0}^{\infty} \frac{t^n H_n(x)^2}{2^n n!} e^{-x^2} = \frac{1}{\sqrt{1 - t^2}} e^{-\frac{1-t}{1+t} x^2}$$

18) $$\sum_{n=0}^{n} H_{n+k}(x) \frac{t^n}{n!} = H_k(x - t) e^{2xt-t^2}$$

19) $$H_n(x) e^{-\frac{1}{2} x^2} = \frac{\Gamma(n+1)}{\Gamma(\frac{n}{2}+1)} [\cos(\sqrt{2n+1}\, x - \frac{n}{2}\pi) + O(\frac{1}{\sqrt{n}})] \quad x : \text{fixed}$$

20) $$H_m(x) H_n(x) = \sum_{k=0}^{\min(m,n)} 2^k k! \binom{m}{k} \binom{n}{k} H_{m+n-2k}(x)$$

21) $$2^{\frac{1}{2m}} H_m(x+y) = \sum_{k=0}^{m} \binom{m}{k} H_k(\sqrt{2}\, x) H_{m-k}(\sqrt{2}\, y)$$

22) $$2^{m-1}\{H_{2m}(x+y) + H_{2m}(x-y)\} = \sum_{k=0}^{m} \binom{2m}{2k} H_{2k}(\sqrt{2}\, x) H_{2m-2k}(\sqrt{2}\, y)$$

23) $$F(x, t) \equiv \sum_{m=0}^{\infty} \frac{a_m t^m}{m!} H_m(x)$$

$$\Rightarrow e^{2xt-t^2} F(x-t, ty) = \sum_{0}^{\infty} \frac{t^r}{r!} b_r(g) H_r(x), \text{ where } b_r(y) = \sum_{m=0}^{r} \binom{r}{m} a_m y^m.$$